ALSO BY LEONARD MLODINOW

*The Upright Thinkers: The Human Journey from
Living in Trees to Understanding the Cosmos*

Subliminal: How Your Unconscious Mind Rules Your Behavior

War of the Worldviews (with Deepak Chopra)

The Grand Design (with Stephen Hawking)

The Drunkard's Walk: How Randomness Rules Our Lives

A Briefer History of Time (with Stephen Hawking)

Feynman's Rainbow: A Search for Beauty in Physics and in Life

*Euclid's Window: The Story of Geometry
from Parallel Lines to Hyperspace*

FOR CHILDREN (with Matt Costello)

The Last Dinosaur

Titanic Cat

elastic

elastic

Flexible Thinking in a Time of Change

LEONARD MLODINOW

PANTHEON BOOKS

NEW YORK

Pantheon Books and colophon are registered trademarks
of Penguin Random House LLC.

Grateful acknowledgment is made to Basic Books for permission to
reprint an excerpt from *Perfectly Reasonable Deviations from the
Beaten Track*, edited by Michelle Feynman, copyright © 2005 by
Michelle Feynman. Reprinted by permission of Basic Books, an imprint
of Perseus Books, LLC, a subsidiary of Hachette Book Group, Inc.

Library of Congress Cataloging-in-Publication Data
Name: Mlodinow, Leonard, [date], author.
Title: Elastic : flexible thinking in a time of change / Leonard Mlodinow.
Description: New York : Pantheon Books, 2018.
Identifiers: LCCN 2017015377. ISBN 9781101870921 (hardcover).
ISBN 9781101870938 (ebook). ISBN 9780375715242 (export edition).
Subjects: LCSH: Neurosciences—Research.
Neurology—Technological innovations. Decision making.
BISAC: SCIENCE / Life Sciences / Neuroscience.
BUSINESS & ECONOMICS / Decision-Making & Problem Solving.
Classification: LCC RC337 .M57 2018. DDC 612.8072—dc23.
LC record available at lccn.loc.gov/2017015377.

www.pantheonbooks.com

Jacket design by Tyler Comrie

Printed in the United States of America
First Edition
2   4   6   8   9   7   5   3   1

For Donna Scott

Contents

Contents

Contents

elastic

Introduction

The Demands of Change

On July 6, 2016, Niantic, a forty-person startup company founded by ex-employees of Google's "Geo" division, launched *Pokémon Go*, an "augmented reality" game that employs a phone's camera to let people capture virtual creatures that appear on their screens as if they exist in the real world. Within two days the app had been installed on more than 10 percent of all Android phones in the United States, and within two weeks it had thirty million users. Soon iPhone owners were spending more time each day on *Pokémon Go* than on Facebook, Snapchat, Instagram, or Twitter. Even more impressive, within days of the game's release, the words *Pokémon Go* drew more searches on Google than the word *porn*.

If you're not a gamer, you might roll your eyes or shrug at all that, but in the business world, the events were hard to ignore: The game generated an astonishing $1.6 million in revenue each day from domestic Apple users alone. Just as important, it added $7.5 billion to Niantic's market value virtually overnight, and within a month it had doubled the stock price of Nintendo, the company that owns the Pokémon trademark.

In its first six months of existence, more than six hundred million people downloaded the *Pokémon Go* app. Contrast that with some of the greatest successes of the early 2000s. Facebook launched in 2004, but it didn't hit the thirty-million-user mark until 2007. The hugely popular *World of Warcraft* game, also released in 2004, took six years to climb to its peak of twelve million subscribers. What seemed like pedal-to-the-metal growth back then became, ten years later, life in the slow lane. And though no one can predict what the next big new thing will be, most economists and sociologists expect that society will only continue to morph faster in the foreseeable future.

But to focus only on the speed of *Pokémon Go*'s ascent is to miss much of the point. The game's massive success might not have been predictable, but neither was it accidental. In creating the app, Niantic made a series of innovative and forward-thinking decisions concerning the use of technology, such as piggybacking on the GPS and camera capabilities of a cell phone and leveraging cloud computing to power the app, which provided a built-in infrastructure and a capacity to scale. The game also took advantage, like nothing before it, of app-store economics, a business model that hadn't even been invented when *World of Warcraft* launched. In that now familiar approach, a game is given away free of charge and makes its money by selling add-ons and upgrades. Maintaining that revenue stream was another challenge. In the interactive entertainment industry, a game can start out popular and still have the shelf life of raw oysters. To avoid that fate, Niantic surprised many with a long campaign to aggressively update the app with meaningful features and content. As a result, a year after its launch, 65 million people were still playing the game each month, and revenues had reached $1.2 billion.

Before *Pokémon Go*, the conventional wisdom was that people didn't want a game that required physical activity and real-world interaction. And so, despite all the innovation in Silicon Valley, the *Pokémon Go* developers were often admonished that gamers just "want to sit and play." But the developers ignored that widely held assumption, and by leveraging existing technologies in a novel way, they changed the way game developers think. The flip side of the *Pokémon Go* story is that if your thinking is not deft, your company can quickly sink. Just look at BlackBerry, Blockbuster, Borders, Dell, Eastman Kodak, Encyclopaedia Britannica, Sun Microsystems, Sears, and Yahoo. And they are just the tip of the iceberg—in 1958, the average life span of companies in the S&P 500 was sixty-one years. Today it is about twenty.

We have to face analogous intellectual challenges in our daily lives. Today we consume, on average, a staggering 100,000 words of new information each day from various media—the equivalent of a three-hundred-page book. That's compared with about 28,000 a few decades ago. Due to innovative new products and technologies, and to that proliferation of information, accomplishing what was once a relatively straightforward task can now be a bewilderingly complex journey through a jungle of possibilities.

Not long ago, if we wanted to take a trip, we'd check out a guidebook or two, get AAA maps, and call the airline and hotels, or we'd talk to one of this country's eighteen thousand travel agents. Today, people use, on average, twenty-six websites when planning a vacation, and must weigh an avalanche of offers and alternatives, with prices that not only change as a function of when in the day you wish to travel but also as a function of when you are *looking*. Simply finalizing the purchase once you've decided has become a kind of duel between business and customer, with each vying for the best deal, from his or her vantage point. If you didn't need a vacation when you started planning one, you might by the time you are done.

Today, as individuals, we have great power at our fingertips, but we must also routinely solve problems that we didn't have to face ten or twenty years ago. For instance, once, while my wife and I were out of the country, my daughter Olivia, then fifteen, gave the house sitter the night off. Olivia then texted us asking if she could invite "a few" friends over. "A few" turned out to be 363—thanks to the instant invitations that can be communicated over cell phones on Instagram. As it turned out, she wasn't entirely to blame—it was an overzealous friend who posted it—but it's a calamity that wouldn't have been possible when her brothers were that age, just a handful of years earlier.

In a society in which even basic functions are being transformed, the challenges can be daunting. Today many of us must invent new structures for our personal lives that account for the fact that digital technology makes us constantly available to our employers. We must discover ways to dodge increasingly sophisticated attempts at cyber-crime or identity theft. We have to manage ever-dwindling "free" time so that we can interact with friends and family, read, exercise, or just relax. We must learn to troubleshoot problems with home software, phones, and computers. Everywhere we turn, and every day, we are faced with circumstances and issues that would not have confronted us just a decade or two ago.

Much has been written about that accelerating pace of change and the globalization and rapid technological innovation that have fueled it. This book is about what is not so often discussed: the new demands on how we must *think* in order to thrive in this whirlwind era—for as rapid change transforms our business, professional, political, and personal environments, our success and happiness depend on our coming to terms with it.

There are certain talents that can help us, qualities of thought that have always been useful but are now becoming essential. For example: the capacity to let go of comfortable ideas and become accustomed to ambiguity and contradiction; the capability to rise above conventional mind-sets and to reframe the questions we ask; the ability to abandon our ingrained assumptions and open ourselves to new paradigms; the propensity to rely on imagination as much as on logic and to generate and integrate a wide variety of ideas; and the willingness to experiment and be tolerant of failure. That's a diverse bouquet of talents, but as psychologists and neuroscientists have elucidated the brain processes behind them, those talents have been revealed as different aspects of a coherent cognitive style. I call it *elastic thinking*.

Elastic thinking is what endows us with the ability to solve novel problems and to overcome the neural and psychological barriers that can impede us from looking beyond the existing order. In the coming pages, we will examine the great strides scientists have recently made in understanding how our brains produce elastic thinking, and how we can nurture it.

6

In that large body of research one quality stands out above all the others—unlike analytical reasoning, elastic thinking arises from what scientists call "bottom-up" processes. A brain can do mental calculations the way a computer does, from the top down, with the brain's high-level executive structures dictating the approach. But, due to its unique architecture, a biological brain can also perform calculations from the bottom up. In the bottom-up mode of processing, individual neurons fire in complex fashion without direction from an executive, and with valuable input from the brain's emotional centers (as we'll be discussing). That kind of processing is nonlinear and can produce ideas that seem far afield, and that would not have arisen in the step-by-step progression of analytical thinking.

Though no computer and few animals excel at elastic thinking, that ability is built into the human brain. That's why the creators of *Pokémon Go* were able to quiet the executive functions of their brains, look beyond the "obvious," and explore entirely new avenues. The more we understand elastic thinking and the bottom-up mechanisms through which our mind produces it, the better we can all learn to harness it to face challenges in our personal lives and our work environments. The purpose of this book is to examine those mental processes, the psychological factors that affect them, and, most important of all, the practical strategies that can help us master them.

Rising Above the Nematode

Every animal has a toolbox for handling the circumstances of daily life, with some capacity to confront change. Take the lowly nematode, or roundworm (*C. elegans*), one of the most primitive biological information-processing systems we know. The nematode either solves its problems of existence by employing a neural network composed of a mere 302 neurons, with only five thousand chemical synapses between them, or it perishes.

Perhaps the most critical challenge the nematode experiences arises when its environment runs out of the microbes it feeds on. Upon recognizing that circumstance, what does this biological computer do?

It crawls into the gut of a slug, waiting to be pooped out the next day in a different location. Not a very glamorous life. To us, the plan may sound both brilliant and disgusting, but in the roundworm's world it is neither, for the few hundred neurons in its nervous system are incapable of either complex problem-solving or sophisticated emotions. To hitchhike in slug excrement is not a desperate creation of the nematode's mind. It is an evolutionary response to deprivation that is hardwired into each individual, because the depletion of food is an environmental circumstance that such organisms face regularly.

Even among more complex animals, much of an organism's behavior is "scripted," by which I mean it is preprogrammed or automatic, and initiated by some trigger in the environment. Consider the brooding goose, with her sophisticated brain, sitting on her nest. When she notices that an egg has fallen out, she fixates on the stray egg, raises herself, and extends her neck and bill to gently roll the egg back into her nest. Those actions appear to be the product of a thoughtful and caring mother, but, like the nematode's, they are simply the product of a script.

Scripted behavior is one of nature's shortcuts, a reliable coping mechanism that leads to results that are usually successful. It can be either innate or the result of habit, and it is often related to mating, nesting, and killing prey. But—what is most important—while scripted behavior can be appropriate in routine situations, it produces a fixed response, and so it often fails in circumstances of novelty or change.

Suppose, for example, that as the goose begins to extend her neck, the stray egg is removed. Will she adapt and abort her plan of action? No, she will continue as if the egg were still there. Like a mime, she will nudge the now imaginary egg back toward her nest. What's more, she can also be induced to perform her egg rolling on any roundish object, such as a beer can or a baseball. In the wisdom of evolution, it was apparently more efficient to endow the mother goose with an automatic behavior that is *almost* always appropriate than to leave the egg-saving action to some more complex but nuanced mental process.

Humans follow scripts, too. I like to think that I give more thought to my actions than the average mother goose. Yet I've found myself, when passing the snack cabinet, grabbing a handful of almonds without pondering whether, at that moment, I really wanted a snack. When my daughter asks if she can stay home from school because she feels a cold "coming on," I may respond with an automatic "No" instead of taking the request seriously and asking for specifics. And I've found myself, while driving to a familiar place, following my familiar route without making a conscious decision to do so.

Scripts are useful shortcuts, but for most animals it would be difficult to survive by employing preprogrammed scripts alone. After recognizing her prey from a distance, for example, a hunting female lion must carefully stalk her quarry. The environment, the conditions, and the actions of her prey can vary considerably. As a result, no *fixed* script inscribed in her nervous system will be adequate to meet the demands of finding food. Instead, the lion must have the ability to evaluate a situation in the context of a goal and to formulate a plan of action aimed at achieving that goal.

It is for those situations in which scripted modes of information processing do not serve an individual well that evolution has provided the two other means through which we and other animals can calculate a response. One is rational/logical/analytical thought, which, for simplicity, I will simply call *analytical thought*—a step-by-step approach through which an organism moves from one related thought to another based on facts or reason. The other is elastic thinking. Different species possess these in differing degrees, but they are thought to be most developed in mammals, especially in primates; and among primates, especially in humans.

Analytical thought is the form of reflection that has been most prized in modern society. Best suited to analyzing life's more straightforward issues, it is the kind of thinking we focus on in our schools. We quantify our ability in it through IQ tests and college entrance examinations, and we seek it in our employees. But although analytical thinking is powerful, like scripted processing it proceeds in a linear fashion. Governed by our conscious mind, in analytical thinking, thoughts and ideas come in sequence, from A to B to C, each following its predecessor according to a fixed set of rules—the rules of logic, as might be executed on a computer. As a result, analytical reasoning, like scripted processing, often fails to meet the challenges of novelty and change.

It is in meeting those challenges that elastic thinking excels. The process of elastic thought cannot be traced in an A to B to C fashion. Instead, proceeding largely in the unconscious, elastic thinking is a nonlinear mode of processing in which multiple threads of thought may be pursued in parallel. Conclusions are reached from the bottom up through the minute interactions of billions of networked neurons in a process too complex to be detailed step by step. Lacking the strict top-down direction of analytical thought, and being more emotion-driven, elastic thinking is tailored to integrating diverse information, solving riddles, and finding new approaches to challenging problems. It also allows the consideration of ideas that are unusual or even bizarre, fueling our creativity (which also requires analytical thinking so that we may understand and explore those new ideas).

Our elastic thinking skills evolved hundreds of thousands of years ago so that we could beat the odds presented by living in the wild. We needed those skills because, as primates go, we aren't the toughest physical specimens. Our close relative the bonobo can jump twice as high. The chimpanzee has, pound for pound, twice the arm strength. A gorilla might find a sharp-angled boulder, have a seat, and survey its surroundings; humans sit on posh chairs and wear glasses. And if it's the wrong chair, we complain about a backache. Our ancestors were no doubt tougher than we are today, but what saved us from extinction was our elastic thinking, which gave us the ability to overcome challenges through social cooperation and innovation.

In the past 12,000 years, we humans have settled into societies that are somewhat protected from the dangers of the wild. Over those many millennia, we've turned our powers of elastic thinking toward improving or enhancing our everyday existence. Robins' nests don't have bathrooms, and squirrels don't store their acorns in safes. But we humans live in an environment built almost entirely on our own imagination. We don't just live in generic huts; we have homes and apartments of all designs and sizes, and we decorate them with works of art. We don't just walk or run; we bicycle, drive cars, travel in boats, and fly in airplanes (not to mention scooting along on Razors, Segways, and hoverboards). Each of these modes of travel, at one time, did not exist. They were each, at conception, a never-before-imagined solution to some perceived problem. As were the eraser and paper clips on your desk, the shoes on your feet, and the toothbrush in your bathroom.

Wherever we go, we swim among the products of the elastic human mind. But though elastic thinking is not a new talent for the human species, the demands of this moment in history have thrust it from background to foreground and made it a critical aptitude in even the routine matters of our everyday professional and personal lives. No longer the special tool of people such as scientific problem solvers, inventors, and artists, a talent for elastic thinking is now an important factor in anyone's ability to thrive.

Onward

Psychologists and neuroscientists are only now working out the science of elastic thinking. They have discovered that the brain function that produces bottom-up elastic thinking is quite different from that which generates top-down analytical thinking. That science rests on recent advances in the study of the brain that have recast our understanding of many of its unique and distinct neural networks. For example, in 2016, the NIH's Human Connectome Project, employing revolutionary new high-resolution imaging techniques and cutting-edge computer technology, showed that the brain has far more substructures than anyone had previously believed. One important structure, the dorsolateral prefrontal cortex, was discovered to actually consist of a dozen distinct smaller elements. In all, the project identified ninety-seven new brain regions, differentiated by both structure and function. The lessons of the Connectome project have opened new vistas, and have been compared to the discovery in physics that atoms are made up of smaller particles—protons, neutrons, and electrons. In the chapters ahead, I will make use of such cutting-edge neuroscience and psychology to explore how elastic thinking arises in the brain. Once we understand these bottom-up thought processes, we will learn ways of implementing, manipulating, controlling, and nurturing them.

Part I of *Elastic* is about how we must adapt our thinking to change, and why our brains are good at it. In part II, I examine how humans (and other animals) take in information and process it so that they can innovate to meet the challenges of novelty and change. Part III is about how the brain attacks problems and generates new ideas and solutions, and part IV is about the barriers that can stand in the way of elastic thinking and how we can overcome them.

Along the way, I will examine the psychological factors that are important in elastic thinking and how they manifest in our lives. These include personality traits such as neophilia (the degree of affinity for novelty) and schizotypy (a cluster of characteristics that include a tendency to have unusual ideas and magical beliefs). They also include abilities such as pattern recognition, idea generation, divergent thinking (being able to think of many diverse ideas), fluency (being able to quickly generate ideas), imagination (being able to conceive of what does not exist), and integrative thinking (the ability to hold in mind, balance, and reconcile diverse or opposing ideas). Research on the brain's role in these traits constitutes one of the hottest new directions in both psychology and neuroscience.

How do our minds respond to the demands of novelty and change? How do we create new concepts and paradigms, and how can we cultivate that ability? What keeps us tied to the old ideas? How can we become flexible in the way we frame questions and issues? We are fortunate that today the enormous mountain of new scientific knowledge about how the bottom-up mind works makes it possible to answer such questions. As I unpack the science of the bottom-up thought mechanisms behind elastic thinking, I hope to change the way you view your own thought processes, and to provide insight into how we think—and how we can think better—so that we can succeed in a world in which an ability to adapt is more crucial than ever before.

Part I

Confronting Change

1

The Joy of Change

The Peril and the Promise

In the early days of television, a program called *The Twilight Zone* had an episode in which a race of nine-foot-tall aliens called the Kanamits land on Earth. They speak an unknown language, but they are able, via telepathy, to address the United Nations, where they vow that their sole purpose in coming is to aid humanity. They give the humans a book in their language, and cryptographers soon decode the title as *To Serve Man*, but they can't make out the meaning of the text within.

In time, with Kanamit technology, deserts are transformed into fertile green fields, and poverty and hunger disappear. Some lucky individuals are even allowed to volunteer for a trip to see the Kanamit planet, said to be a paradise. And then one of the cryptographers finally breaks the code. She reads *To Serve Man*, then races to the ship, where her boss, a fellow named Michael Chambers, is on the steps leading up to the entrance, about to depart for the alien planet himself. "Don't get on!" she yells to Chambers. "The book is a cookbook!" A cookbook in which *humans* are the main ingredient.

The cryptographer had discovered that the aliens were here to help us, but in the manner that farmers help turkeys in the days before Thanksgiving. And apparently, having a sense of humor, they left us a book of the recipes they planned to use. Chambers tries to disembark, but there is one of those nine-foot-tall aliens beside him. Not wanting to lose a yummy tidbit for his human stew, the alien blocks Chambers's retreat.

The obvious moral of the Kanamit story is that there's no such thing as a free lunch—unless you're the lunch. But it is also about the peril and the promise of novelty and change. When an animal ventures out into new territory, it could lead to the discovery of a food source—or to becoming one. A novelty-seeking organism might get injured exploring alien terrain or might face a predator, but an organism that avoids the unfamiliar at all costs might fail to discover sufficient food sources and starve.

An unchanging environment offers those who have found a comfortable niche no urgent impetus to explore or innovate. But conditions do change, and animals have a better shot at survival if they have previously gathered information about new feeding sites, escape routes, hiding places, and so on. Biologists see that reflected in the varying character of species. For example, dogs like to explore new territory because they're descended from particularly daring wolves that ventured out looking for food around the campsites of ancient human nomads; and birds that live in a complex and changing habitat, such as the edge of the forest, tend to exhibit more exploratory behavior than those that live in less variable surroundings.

Today it is we humans who must adapt, for our physical, social, and intellectual environments are changing at an unparalleled pace. Scientific knowledge, for example, grows exponentially—that is, the number of published papers doubles at a fixed rate, just as money invested at a fixed interest rate does. In the case of global scientific output, the doubling occurs about every nine years. That has been the case for a long time, but in the past that growth was possible to assimilate because if you don't start with much in the first place, doubling does not represent a large incremental increase. Today, however, the volume of our knowledge has exceeded an important milestone. Today, to double our knowledge each nine years means to add new knowledge so fast that no human can keep up. In 2017, for example, there were more than three million new scientific papers. That rate of production isn't just greater than the practitioners in any given field can assimilate; it is greater than the journals can contain. As a result, in the decade between 2004 and 2014, publishers had to create more than five thousand new scientific journals just to accommodate the overflow.

In the professional world, due to an analogous expansion of knowledge, many major industries also now depend on a volume of expertise that no single person could ever master. Arcane topics from electric transformers to fuel injection to the chemistry of cosmetics and hair products are each the subject of *hundreds* of books—and that doesn't include the proprietary knowledge held by the corporations in those businesses. The intricacies of "fuzzy logic optimization of injection molding of liquid silicone rubber" may not interest you, but it is an important enough subject in today's world that Firmin Z. Sillo wrote a 190-page book on it in 2005.

The growth of social media and the Internet is even more drastic: the number of websites, for example, has been doubling every two to three years. Social attitudes, too, are changing fast—just compare the pace of the civil rights movement to the speed at which the campaign for gay rights has swept the developed world, again fueled by the young.

There is peril and promise in every decision about whether or not to embrace novelty. But in the recent past, as the pace of change has quickened, the calculus governing the benefits of embracing novelty has been dramatically altered. Today's society bestows rewards as never before upon those who are comfortable with change, and it may punish those who are not, for what used to be the safe terrain of stability is now often a dangerous minefield of stagnation.

Consider the history of the telephone. We use the phrase "dial a number" because telephone numbers used to be input by turning a numbered dial. A new technology, push-button dialing, was introduced by Bell Telephone in 1963. It was more convenient than the older system and offered the possibility of making menu choices in response to automated phone systems. But the technology was not a great investment, at least in the short term, because people were slow to alter their habits and adopt it, preferring to stick to the comfortable phones of the past. Even twenty years after the "touch-tone" devices became available, the majority of customers still had the old "rotary" phones. It wasn't until the 1990s, three decades after the introduction of touch-tone phones, that the older type of telephone became a rarity.

Contrast that with what happened when Apple, in 2007, introduced the first practical touchscreen mobile phone, meant to replace the existing keypad or stylus phones. The Apple iPhones were immediately the rage, and within several years the competing technologies virtually vanished. Unlike the prior age, in which adoption proceeded at a snail's pace, in 2007 people were not just ready but eager to change their habits, and hungry for every new phone version and capability that emerged in the years that followed.

In the mid-twentieth century it took decades for people to change their simple habit of using a dial phone, while in the twenty-first century it took very little time for people to make the transition to carrying around with them what is essentially an entire computer system. Companies like BlackBerry that didn't immediately adapt to the new technology were quickly marginalized, but adaptation soon became equally important for individuals to reach their potential and to thrive socially.

The Kanamits episode of *The Twilight Zone* aired just a year before the introduction of the touch-tone phone. At the end of the episode, Chambers, now on the ship, turns to the camera and asks the viewers, "How about you? You still on Earth, or on the ship with me?" The implication was that it could be deadly to go along with what is new or different. Today, when alien ideas land in your professional or social world, it's a better bet to take your chances, climb aboard the spaceship, and check them out.

The Myth of Change Aversion

Would you have climbed aboard the Kanamit ship? A widespread myth in our culture holds that people are averse to novelty and change. Change is an issue that arises often in the world of work, and the academic business literature has much to say about it. "Employees tend to instinctively oppose change," proclaimed one article in the *Harvard Business Review*. "Why is change so hard?" asked another. But is change really that hard? If people are generally averse to change, psychologists must have missed it, because if you instead search the literature of research psychology, you'll find nary a mention of change aversion.

The reason for this difference in perception is that, while management endows change initiatives with names like *restructuring, turnaround,* and *strategic shift,* employees often see them as something else: layoffs. When change translates to the risk of losing one's job, or what is novel is an increased workload, it's understandable that people react negatively. But that's not change aversion; it is unemployment aversion or negative consequence aversion.

An employee might bristle at being called into a superior's office to be told, in essence, "The corporation is striving to be more efficient, so you will be asked to do ten percent more work for the same salary." But that same employee would delight at being told, "The corporation is striving to be less efficient, so you will be asked to do ten percent less work for the same salary." That's two opposite reactions to the same degree of change. The latter request never happens, but if it did, those *Harvard Business Review* articles would be saying, "Employees tend to instinctively *love* change," and asking, "Why is change so *easy*?"

To avoid change because it is negative or requires work or introduces the risk of either of those eventualities is a rational and logical reaction. But as far as human nature goes, in the absence of negative consequences, our natural instinct is the opposite: We humans tend to be *attracted* to both novelty and change. That trait, called "neophilia," *is* a topic that is written about in the literature of academic psychology. Indeed, along with reward dependence, harm avoidance, and persistence, neophilia is considered one of the four basic components of human temperament.

An individual's general attitude toward novelty and change is affected by both nature and nurture—our genes and our environment. The influence of our environment is most apparent in the evolution of human attitudes over time. A few centuries ago, most people's lives were characterized by repetitive tasks, long hours of solitude, and a dearth of stimulation. Novelty and change were rare, and people were suspicious of it while being perfectly comfortable with conditions that we would today find extremely tedious. And by "extremely tedious" I don't mean as in the time your girlfriend dragged you to a documentary on the life of Al Gore. I mean as in a sixty-hour workweek spent chipping rough pieces of rock so that they can be stacked to build a structure, or using a hand ax to chop down and trim a fifty-foot maple tree, or spending weeks sitting in a cramped stagecoach while traveling from New York to Ohio.

Because tedium used to be the norm, the concept of "boring"—or at least the English word for it—didn't even appear until the industrial revolution, in the late eighteenth century. Since then, both the availability of stimulation and our thirst for it have gradually grown—especially in the twentieth century, which saw the rise of electricity, radio, television, movies, and new modes of transportation. That not only brought changes in the way we live; it also exposed us to other ways of living, vastly increasing our mobility and the number of new people and places we encountered. Through travel and the media, we could explore not just our own towns and cities but the entire world.

Though in the twentieth century we became much more comfortable with novelty and change, that evolution of our attitudes was nothing compared with the transformation wrought by the advances of the past twenty years, by the rise of the Internet, email, texting, and social media, and the increased pace of technological change.

Our evolving attitude is an adaptation, but it is also a blossoming, for we have always had the potential to make great adjustments. As we'll see, it is in our genes. It is one of our defining traits. We'll get to individual differences later, and to trends that depend on one's genetics, experience, and age, but on the whole, those in the business world who grouse about people's reluctance to adapt to modifications at the workplace are lucky that they don't have to make cats work new hours or raccoons alter the way they forage for food. For, compared with other species, humans *love* novelty and change. "We [humans] jump borders. We push into new territory even when we have resources where we are. Other animals don't do this," says Svante Pääbo, director of the Max Planck Institute for Evolutionary Anthropology.

So although our current era is making unprecedented demands on us, it's actually just asking us to tap into a quality we've had all along—one of the qualities that make us human. Our ability and desire to adapt, to explore, and to generate new ideas are, in fact, what this book is all about.

Our Exploratory Drive

Early versions of our species weren't neophilic. Two hundred thousand years ago in Africa, our ancestors had no apparent drive to probe new environments. The crew of *Star Trek* was on a mission "to explore strange new worlds, to seek out new life and new civilizations, to boldly go where no man has gone before," but a crew with the attitude of the earliest humans would more likely have gone on a mission to "sit on a log, take no chances, and timidly avoid areas that no one has checked out."

What seems to have changed in our psyches was a great catastrophic event—probably related to climate change—that decimated our ranks about 135,000 years ago. At that time, the entire population of the subspecies that we now call human plummeted to just six hundred. Today, that would be low enough to land us on the endangered species list, finally ensuring that the list contains at least one animal that everyone can agree is worth saving. But though the die-off was no doubt a tragic time for most of our ancestors, it was a blessing for those of our species that survived.

Many scientists now believe that that environmental battering acted as a genetic filter, culling from our ranks the less adventurous and preferentially allowing to survive those with the bold desire to explore. In other words, had they lived back then, those friends who always go to the same restaurant and order the steak and potatoes would likely have perished, while the thrill-seekers who revel in discovering new chefs and dishes like rotten shark and fried pig's ear would have had a better chance of enduring.

Scientists drew this conclusion because for hundreds of thousands of years, humans had remained close to their origins in Africa. But then, as fossils discovered in China and Israel reveal, within a few thousand years of the die-off, the descendants of those hardy survivors were "suddenly" traveling to distant new worlds. In 2015, those discoveries were bolstered by analysis of both modern populations and ancient genetic material. These indicate that by fifty thousand years ago, humans had spread throughout Europe, and by twelve thousand years ago, to every corner of the globe. As colonization goes, that was swift, and suggests an evolution in the fundamental character of our species. Neanderthals, by comparison, were around for hundreds of thousands of years, and never spread beyond Europe and central and western Asia.

If our species was altered by that catastrophic event—if that harsh era of our existence favored those with a greater tendency to explore and take chances—then our attitude toward change should be reflected in our genetic makeup. Our species today ought to possess a gene or a set of genes that drives us to be discontented with the status quo, to seek the new and unfamiliar. Scientists found just such a gene in 1996. It is called DRD4, for "dopamine receptor gene D4," because it affects the way the brain responds to dopamine.

Dopamine is a neurotransmitter, one of several protein molecules that neurons use to communicate with one another. It plays an especially important role in the brain's reward system, which I will talk about in chapter 3. For now, I will merely point out that the reward system initiates your feelings of pleasure, and dopamine carries those signals. Without your reward system, you'd feel the same whether it's a traffic cop telling you "I'll let you go this time with a warning" or a CNN reporter saying "Scientists have just discovered exoplanet number 4000."

The DRD4 gene comes in variants named DRD4-2R, DRD4-3R, etc. Everyone has some form of the gene, but just as height and eye color vary, so does the degree of novelty-seeking bestowed by those different forms. Some versions of the gene, such as the DRD4-7R variant, endow people with a particularly high tendency to explore. That's because those with that variant respond more weakly to the dopamine in their reward system. As a result, they require more dopamine to get a rush in their day-to-day life than those with other variants, and seek a higher level of stimulation in order to achieve a satisfying level.

The discovery of the role of DRD4 answered some questions, but it raised others. For example, if that gene is truly related to our tendency to explore, do populations that have wandered far from our African origins possess a higher incidence of DRD4-7R than those who strayed less? If our picture of the origin of our novelty-seeking behavior is correct, one would expect that.

That expectation proved valid. The geographic connection was made first in 1999, then more definitively in a landmark 2011 paper with the cumbersome title "Novelty-Seeking DRD4 Polymorphisms Are Associated with Human Migration Distance Out-of-Africa After Controlling for Neutral Population Gene Structure." Those papers reported that the farther our ancestors migrated from their African roots, the higher the prevalence in that population of the DRD4-7R variant. For example, Jews who migrated to Rome and Germany, a long way from their origin, show a higher proportion of that variant than those who migrated a shorter distance southward to Ethiopia and Yemen.

It is an oversimplification to chalk up anything as complex as a personality trait to a single gene. There are certainly many genes that contribute to a tendency toward novelty and exploration. And the genetic component is only one factor in an equation that must also include a person's life history and current circumstances. Still, the genetic contribution can be traced, and scientists are currently seeking other genes that may be involved, and their function, to complete the picture.

The good news, as we face increasing novelty and accelerating change in human society, is that although the changes are disruptive, most of us have a good dose of neophilia as part of our genetic inheritance. The same traits that saved us 135,000 years ago can still help us today.

Even better news, for us and for our species, is that not only do our genes help us cope with the new society, but our society can also help shape our genes. Cutting-edge research in genomics shows that our traits are not, as previously believed, simply consequences of the DNA that makes up our genes. Instead, our traits depend also on "epigenetics"—the way cells modify our genomic DNA and the proteins tightly bound to that DNA in order to turn genes on or off in response to external circumstances. We have only begun to understand how that works, but epigenetic changes can result from your behavior or habits, and may even be heritable. If that proves to be true, the changes in society that favor a greater aptitude for dealing with novelty could eventually cause adaptive changes in our species.

Personal R&D and the Neophilia Scale

You may remember that a couple of decades ago a fellow named Timothy Treadwell was a media sensation and a Hollywood darling. Leonardo DiCaprio reportedly contributed to his fund-raising entity, as did Pierce Brosnan and corporations like Patagonia. Treadwell was an advocate for Alaskan grizzly bears and a celebrated explorer who lived among them.

Psychologists have a term for people on the far end of the novelty-seeking spectrum. They call them "sensation-seekers." Treadwell was a sensation-seeker. Living in Long Beach, California, before he'd ever been to Alaska, he experimented with drugs, such as a "speedball" of heroin and cocaine that nearly killed him. Another night, tripping on LSD, he jumped off a third-floor balcony and landed on his face, luckily in a soft patch of mud. But after he discovered Alaska and its grizzlies, he traded his drug-soaked quests for adventures in the bear country of Katmai National Park, where he would spend each summer living near the bears and interacting with them.

Weighing a thousand pounds, the bears "can run thirty-five miles per hour" and "jump eleven feet in the air," Treadwell marveled. They can also stalk their prey in virtual silence, and "kill you with one swat." Treadwell boldly and patiently explored bear behavior until he believed he'd found the secret to disarming them: singing to the bears and telling them he loved them. "Animals rule, Timothy conquered," he said. "Come here and try to do what I do—you will die here, [but] I found a way to survive with them." In 2003, not long after making that pronouncement, Treadwell and his girlfriend were both eaten alive.

Some like to break 100 driving their Harley down a country road; others opt for a quiet afternoon reading *A History of the Metal Lawn Chair*. Although an extreme proclivity for adventure and exploration may result in a reduced life expectancy for those—like Treadwell—who possess it, the population's overall chance of survival may be increased by the presence of such "pioneers," because the group stands to benefit from their discovery of new resources. And so our species encompasses a spectrum of individuals, from those fearful of risk to brash adventurers like Treadwell who seem indifferent to fear.

In the wild, novelty-seeking human pioneers explored new terrain or, like Treadwell, the lives of the animals living on it. In the context of how we live today, those who generate unusual and original ideas in science, the arts, or business are motivated by the same kind of drive, applied to a different kind of terrain, and the fruits of their efforts are just as influential in our lives in civilized society as they were when we lived in the wild.

We also explore in our personal lives, risking time and money on activities that might—or might not—pay off. It is our individual version of a corporation's R&D. When you socialize with strangers, you are exploring the possibility of new relationships. When you take a night class to learn a skill you haven't tried before, you're exploring a new hobby. When you go on a job interview even though you are employed, you are exploring a new career move. When you start a new business, you are exploring the world of commerce. When you go on Match.com, you are exploring the landscape of romance.

As with other animals, the amount of resources you invest in personal R&D activities depends upon several factors—your degree of satisfaction with your current "environment," your situation in life, and the degree of your innate human propensity to seek the new. Psychologists have developed several "inventories" to measure a person's novelty-seeking tendency. Below is one of them, an eight-statement test that you can take to assess yours. Just rate each statement on a scale of 1 to 5 and compute your total. Use this key for your ratings:

1 = strongly disagree
2 = disagree
3 = neither agree nor disagree
4 = agree
5 = strongly agree

Here are the statements:
1. ___ I would like to explore strange places.
2. ___ I would like to take off on a trip with no preplanned routes or timetables.
3. ___ I get restless when I spend too much time at home.
4. ___ I prefer friends who are excitingly unpredictable.
5. ___ I like to do frightening things.
6. ___ I would like to try bungee jumping.
7. ___ I like wild parties.
8. ___ I love to have new and exciting experiences, even if they are illegal.

Total: ___

As the graph below illustrates, if your points total 24, you fall in the middle of the population on the neophilia scale. About two-thirds of all individuals score within five points of that—between 19 and 29. People who score particularly high are natural explorers. Those who score low are the ones talented at providing reality checks, stability,

16% fall here	68% fall here	16% fall here
0 10	19 24 29	40
low in neophilia	average	high in neophilia

Distribution of neophilia scores

26

and risk assessment. They may also be more practical. I scored 37, which my mother tells me was predictable, given that at age twelve I jumped off the roof of my school just to see how it felt. (It felt better a few weeks later, which was also when I could walk again.)

Had I taken a neophilia test at age twelve, I'd probably have scored even higher, for as the graph below indicates, the degree to which we are attracted to the new and sensational varies with age. In a study of young adults aged eighteen to twenty-six, the average score was several points higher than the adult average: 27.5. And in a study of adolescents aged thirteen to seventeen, the average was 30, one point above the cutoff I just quoted for extreme novelty-seeking adults.

Certainly, that younger people exhibit greater neophilia is due in part to the more rapidly changing world they are growing up in. But since seeking novelty entails risk, this age variation is also no doubt partly due to the fact that the rational, risk-avoiding part of a person's brain doesn't fully develop until about the age of twenty-five.

While degree of neophilia is an important indicator of your comfort in confronting novelty and change, it is your cognitive style—your manner of drawing conclusions, making decisions, and solving problems—that determines the approach you take when facing the challenges that arise from such situations. Your cognitive style is probably neither purely analytical nor purely elastic but,

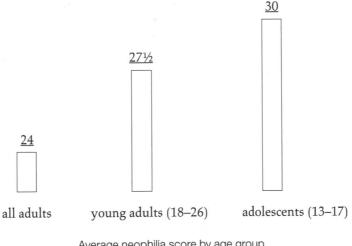

Average neophilia score by age group

rather, has elements of each. And—within bounds that vary among individuals—the mix that you employ will depend upon the situation, your mood, and other factors. Most important, the approach that your mind tends to adopt can be altered if you work at it. The first step in learning to take charge of your thinking is to understand what it means to think, how elastic thinking differs from analytical thought and programmed processing, what drives our thought processes, and how our brains process information. Those are the topics we will address as we move forward to part II.

Part II

How We Think

2

What Is Thought?

Peeking Inside the Skull

Anne Greene, a "fat, fleshy woman," was escorted to the gallows in Oxford, England, on a cold, rainy day in 1650, still proclaiming her innocence. Doctors supported her claim: They believed that her infant had been born too small to survive, so they doubted that she had purposely caused its death, as was the charge. But the baby's father, who accused her of the crime, was the grandson of a powerful local gentleman, and so the judges sentenced her to hang. She climbed the ladder. A psalm was sung. The noose was fitted around her neck, and she was shoved off the ladder.

Anne Greene hung in front of the courtyard crowd for a half hour before she was pronounced dead and finally cut down. She was placed in a coffin provided by Drs. Thomas Willis and William Petty, physicians who had permission from King Charles I to dissect the bodies of criminals for the purpose of medical research. The coffin was carried to a dissecting room in Petty's house, where the autopsy was to be performed. But when they opened the coffin to remove and cut into the corpse, Petty heard a grumbling in Greene's throat.

After years of cutting up the dead, this was the first time Petty had had a corpse protest. He felt her neck and found a faint pulse. The two doctors stood over the body and rubbed the woman's hands and feet for fifteen minutes. They dabbed turpentine over the burn on her neck, then tickled the inside of her throat with a feather. Sounds like a *Saturday Night Live* skit, but it worked. She coughed. The next morning, Anne Greene felt alive again. She asked for a beer. A few days later she was out of bed and "eating chicken wings."

The authorities decided to hang her again, revealing themselves to be somewhere between Lord Voldemort and Josef Mengele on the mercy scale. But Willis and Petty argued that Anne Greene's survival was a sign of divine providence indicating her innocence, and Greene was eventually set free. She went on to marry and have several more children.

Before moving out of Petty's house, Greene was able to make some money by returning to her coffin—people paid to file past and gaze at her lying there, the woman who had come back from the threshold of dissection. The incident also conferred fame and prestige on Thomas Willis, who had "brought her back from the dead" and let everybody know it. Poets wrote works in his honor, and he became one of the best-known doctors of his time.

In his dissections, Willis had focused on the brain. By performing autopsies on patients he had treated throughout their lives, Willis was able to study the connection between brain damage and abnormal behavior. He became the first to connect such behavior to specific changes in brain structure. He coined the term *neurology* and identified and named many of the regions of the brain that we still study today. He used his newfound fame to publish and publicize his work and his ideas. And with help from the architect Christopher Wren, he created drawings of the human brain that for the next two centuries were the most accurate available.

Three hundred years after Willis's death, we no longer have to wait until people are dead to look inside their heads. Technology has given us the means to study brains while they are still alive, and helped spawn the new field of cognitive neuroscience—the study of how we think, and how that thinking is produced by the brain.

It is one of the basic tenets of cognitive neuroscience that the structure and form of thought is independent of its specific content. In other words, the mental activity that leads to the creation of new businesses, shampoos, and meals is fundamentally the same as that which produces new scientific theories, paintings, and symphonies. And so, as we begin to investigate elastic thinking, we are able to consider first, more generally, the nature of thought itself.

What Qualifies as Thought

Why did animals evolve brains? The philosopher Karl Popper addressed this question indirectly when he wrote that "all life is problem solving." Popper's words reflect the perspective of evolutionary biology, which looks at animals as biological machines seeking to survive and reproduce. In that view, animals are seen as contraptions that navigate from challenge to challenge. The evolution of animal brains, then, is the development, over eons, of ever better problem-solving machines. Stepping forward with your foot is solving the problem of getting from here to there, but writing a poem or creating a painting is also solving a problem—that of expressing oneself on some subject or feeling. That's a perspective on thinking that is shared by many neuroscientists and psychologists.

Whether or not *all* of life is problem-solving, it is hard to dispute that, at least in the animal kingdom, a great deal of it is, because it has to be. A rock resting on a hillside makes no effort to alter its destiny. Plants are alive, but they can't do much better. Being stationary, relative to animals, they have less need to confront change, but also less ability. They lay down roots that more or less determine their environment, and they cope with what that entails—or die. Animals, on the other hand, are built to change their circumstances by moving away from threatening conditions and situations and toward favorable ones. That is a useful ability, but because their life involves motion, they must continually act to solve various problems and riddles that they encounter. They accomplish that through senses that gather data, or some other means of detecting what is happening in the environment, and a brain, or brain-like structure, that processes the sensory information so that they can interpret dynamic situations and choose the appropriate action.

But evolution is economical, and it does not create a Maserati where a motor scooter would do. Hence, to solve their problems, animals possess the three increasingly sophisticated modes of information processing I mentioned earlier: scripted, analytical, and elastic. The former addresses simple and routine problems, while other challenges are met through the latter two.

That suggests an interesting question: If an organism is processing information, does that mean it is thinking? Slime mold, a lowly amoeboid, when placed in a maze, will figure out how to propel itself to the food. And if that food is placed at two different sites within the maze, it will re-form itself to engulf them both, in the most efficient manner possible, by morphing into the shortest shape that can reach both places. The slime mold is solving a problem. Is that thinking? And if we don't call it thinking, why doesn't it qualify? Where do we draw the line?

According to the dictionary, to think is "to employ one's mind rationally and objectively in evaluating or dealing with a given situation; to consider something as a possible action, choice, etc.; to invent or conceive of something." A textbook on neuroscience puts it a bit more technically: "Thought is the act of attending to, identifying, and making meaningful responses to stimuli . . . characterized by the ability to generate strings of ideas, many of which are novel."

At their simplest, these definitions say that thinking is *evaluating* circumstances and making a meaningful response by *generating ideas*. That means that scripted information processing, such as that performed by the slime mold, does not qualify as "thinking." The mold is not evaluating a circumstance, but responding to an environmental trigger. It is not generating an idea, but following a preprogrammed response. The same is true of the mother goose, protecting her eggs in the nest.

That said, to exclude from one's definition of *thinking* the fully automatic execution of a script in an organism's (or computer's) programming is just a convention, an arbitrary line we've chosen to draw. What is important to recognize is that, given that definition, what we call thinking is not necessary for much, or most, of an animal's existence. Thinking, in the animal kingdom, is the exception, not the rule, because most animals live largely standard-issue lives. They do just fine, most of the time, acting as automatons. What about us humans? Are our responses the result of thought, or do we, too, go through much of life by scripted habit, without thinking?

Becoming Mindful

In the late 1970s, psychologist Ellen Langer and two colleagues wrote a groundbreaking paper that asked the question "How much behavior can go on without full awareness?" Plenty, they concluded, as was reflected in the title of the paper, "The Mindlessness of Ostensibly Thoughtful Action."

We all know that we sometimes execute actions on "autopilot." But what was shocking in Langer's paper was that such scripted behavior is also common in our "complex social interactions." By "complex" Langer didn't mean drama or Machiavellian plotting. She simply meant an interaction in which something, even something minor, was at stake. When we are confronted with familiar situations of that sort, she and her colleagues concluded, we tend to behave mindlessly, according to programmed patterns, and with relatively little adjustment due to the specifics of the situation at hand.

In one experiment described in the paper, a researcher sat at a table that had a view of a Xerox machine and approached people who walked up intending to make copies. He said, "Excuse me, I have five pages. May I use the machine?" Sixty percent of the Xerox users allowed it. But to others, the researcher instead said, "Excuse me, I have five pages. May I use the machine *because I'm in a rush?*" When asked that way, 94 percent acceded to the request.

As with the mother goose, this appears to be thoughtful behavior. It seems as if most of those who would have been among the 40 percent who refused the first request responded differently when a justification was offered that allowed them to weigh the urgency of their need versus that of the person "in a rush."

But the experimenter also tried out a third version of the request, asking, "Excuse me, I have five pages. May I use the machine *because I have some copies to make?*" This version of the request appears to have the same structure as the successful one: statement, request, justification. But the content differs. This time the "justification" is empty. The phrase "because I have some copies to make" adds no information at all to the prior statement, "I have five pages."

If the Xeroxers were truly deciding how to respond based on the merits of the request, this last approach should have had the same success rate as when no reason was given—60 percent. But if they were following a script that says, "If the requestor offers a reason—a because statement (no matter how irrelevant)—agree to the request," then you'd expect a success rate closer to the other case, 94 percent. Which is exactly what happened: The dummy reason had a success rate of 93 percent. Those who were swayed by the dummy reason were apparently following a mindless script.

This and other research suggests that although you may think you rarely follow scripts in your own social interactions, most of us do it quite often. In fact, clinical psychologists, who work outside the world of controlled laboratory studies, see scripted behavior all the time, especially in the dynamics of relationships. For instance, relationship researchers have identified a pattern called "demand/withdraw" that some couples engage in regularly, even though it is destructive. That dynamic occurs when one partner, typically the woman, seeks a change in the other or a discussion of an interpersonal issue. That's the "demand." It triggers an automatic withdrawal response in many men, who seek to avoid that discussion. If her partner's withdrawal, in turn, triggers the woman to amplify her request, the result can be an escalating conflict.

Analogously, one partner in a relationship may do something to irritate an emotional "raw spot" in his or her counterpart, triggering an angry but predictable reaction. Unfortunately, that anger often serves as a trigger for a reaction in the first partner, who takes the anger personally rather than seeing it as a mindless reaction based on an automated script. The result, again, is an escalation and a familiar cycle of conflict and argument.

Therapists tell their patients that the way out of such cycles is to learn to recognize when they are occurring and then stand together to interrupt the scripts—just as the people standing at the Xerox machine could have, had they been aware of the automatic nature of their reactions. That's analogous to the simple control you exercise when, while driving to work, you hear the siren of an ambulance or encounter some other anomalous circumstance and disengage the autopilot mode in which you usually operate.

More generally, the first step in nurturing either analytical or elastic thinking is to nurture *thinking*—to become more conscious of when you employ automated scripts, and to discard them when they don't serve you well. For it is only when you are self-aware that you can interrupt an automatic script if it is not appropriate. Langer called that self-awareness *wakefulness*. Today, psychologists call it *mindfulness*, building on a concept with roots in Buddhist meditation.

William James said, "Compared with what we ought to be, we are only half awake." A mindful state stands in contrast to that. When you are being mindful, you are fully aware of all your current perceptions, sensations, feelings, and thought processes and accept them calmly, as though seen from a distance. The mental monitoring required is not difficult, but, like improving your posture, it requires continual effort. Luckily, a lot of recent research shows that mindfulness can be cultivated through simple mental exercises. I describe a few of the more well-known exercises below, for those who are interested in trying one.

1. *The Body Scan.* Sit or lie down, in a comfortable position. The activity should take ten to twenty minutes. Loosen any tight clothing and close your eyes. Take a few deep breaths and focus on your body as a whole. Feel its weight on the

floor or chair, and how that contact feels. Then, beginning with your feet, become aware of how each part of your body feels. Are your feet warm or cool, tense or relaxed? Do you feel any sensations, discomfort, or pain? Slowly, let your attention drift to your ankles, calves, knees, thighs, buttocks, and hips, and then up your torso. Next, focus on your fingers, then move up your arms to your shoulders, and finally to your neck, face, head, and scalp. Finally, reverse the process, moving down your body.

2. *Mindfulness of Thoughts.* Like the body scan, this can be done in twenty minutes or less, and you begin by closing your eyes and taking deep breaths. Focus on your breathing until you've quieted your mind. Then relax your concentration and let thoughts drift in. Pay attention to each thought in a detached manner, without judgment or engagement: Is it a feeling, a mental image, a bit of internal dialogue? Does it simply fade, or lead to another thought? If you encounter the thought that you are having difficulty in this exercise, accept and observe that thought as well.

3. *Mindful Eating.* This exercise is shorter, and fun—it should take five minutes. You can perform it with whatever food you like. It is often done with raisins, but I use it as an excuse to eat a piece of chocolate. I'll describe how I do it. Begin, as in the other exercises, by taking a few deep breaths and clearing your mind. Then take the chocolate in your hands. Focus on it. If it is wrapped, feel the wrapper. Turn it in your fingers and feel its texture. Then unwrap it and feel the chocolate. Note its appearance. Bring it to your nose and smell its fragrance. Note how your body reacts to it. Now slowly bring it to your lips and place it gently in your mouth, but don't chew or swallow it. Close your eyes and move your tongue over the chocolate. Pay attention to its feel. Focus on the tastes and sensations you perceive on your tongue. Move the chocolate around in your mouth. Be aware of a desire to

swallow it, if that arises. As it melts, slowly swallow it, staying aware of the sensations.

There are many other mindfulness exercises—you can easily find them on the Internet. Which ones you do doesn't matter, but according to the research, if you perform your exercise of choice three to six times a week, after a month you will have achieved a measurable improvement in your ability to avoid automatic responses, as well as in other "executive functions" of your brain (see chapter 4), such as the ability to focus and to switch your attention from one task to another. Such skills will enable you to exercise more control over how your mind operates, and can bring perspective to the issues and problems that arise in your life.

The Laws of Thought

Once we rise above fixed scripts, the next category of thought is analytical thought. We tend to praise analytical thought as being objective, untinged by the distortions of human feelings, and therefore tending toward accuracy. But though many praise analytical thought for its detachment from emotion, one could also criticize it as not being *inspired* by emotion, as elastic thinking is.

The relative lack of an emotional component is one reason analytical thought is simpler than elastic thought, and easier to analyze. Our first modern insight into its nature came more than a century and a half ago, when, in 1851, the dean of faculty at Queen's College Cork, in southwestern Ireland, gave the annual address for the start of the college session. In that address he asked

> whether there exist, with reference to our mental faculties, such general laws as are necessary to constitute a science . . . I reply that this is possible, and that [the laws of reason] constitute the true basis of mathematics. I speak here not of the mathematics of number and quantity alone, but a mathematics in its larger, and I

believe, truer sense, as universal reasoning expressed in symbolic forms.

Three years later, that dean, mathematician George Boole, published a more elaborate analysis in a book entitled *The Laws of Thought*.

Boole's idea was to reduce logical reasoning to a set of rules comparable to those of algebra. He wasn't completely successful in delivering on the promise of the title, but he did create a way of expressing simple thoughts or statements that allows them to be written as equations that can be combined and operated upon, in a manner analogous to the way addition and multiplication allow us to operate on and form equations involving numbers.

Boole's work grew in importance a hundred years after his death, with the invention of digital computers, which in their early days were called "thinking machines." Today's computers are essentially an implementation in silicon of Boole's algebra, containing circuit elements called "gates" that can string together billions of such logic operations each second.

Boole's farsightedness wasn't confined to mathematics: In the 1830s he became an officer of an organization that advocated putting reasonable legal limits on working hours, and he co-founded a center for the rehabilitation of wayward women. He died in the late fall of 1864. The end came after he took a long walk in torrential rain and then, soaked head to toe, gave a lecture, after which he walked home in the rain. Once back, he collapsed in bed with a high fever. His wife, following the dictates of homeopathy, proceeded to pour bucket after bucket of cold water over him. He died two weeks later, of pneumonia.

Around the time Boole was inventing the mathematics of thought, his fellow Englishman Charles Babbage was trying to build a machine to implement that thinking. Babbage's machine was to be built from thousands of cylinders coupled in a complex manner through intricate gears. He worked on that "Analytical Engine" for decades, beginning in the late 1830s, but because of its complexity and expense, he never completed it. He died in 1871, bitterly disappointed.

Babbage had envisioned the engine as having four main components. The *input*, to come via punched cards, was the mechanism for feeding the machine data as well as instructing it on how to manipulate that data—what we today call the machine's program. The *store* was what Babbage called the machine's memory, analogous to the computer's hard drive. The *mill* was the part of the engine that processed the data according to the instructions that were input—in other words, the central processing unit. The mill also had a small memory, just enough to hold the data being immediately worked on—what we would call random-access memory, or RAM. And finally, there was the *output*, an apparatus to print the answers.

All told, Babbage's machine embodied almost every major principle of the modern digital computer, and, on a superficial level, it offered a new framework for understanding how our minds work. For our brains also have a data input module (our senses), a processing unit for operating on or "thinking" about the data (the cerebral cortex), and both a short-term working memory in which we hold the thoughts or words we are currently considering and a long-term memory for knowledge and rote procedures.

A friend of Babbage's, mathematician Lady Ada Lovelace—the daughter of Lord Byron and his wife, Anne Isabella Noel—wrote that his Analytical Engine "weaves algebraic patterns just as the Jacquard loom weaves flowers and leaves." It was a vivid comparison, though she was jumping the gun, because Babbage hadn't built his machine. Still, Lady Lovelace appreciated the attempt, perhaps even more than Babbage himself. For while he dreamed of his machine playing chess, she saw it as mechanized intelligence, a device that might someday "compose elaborate and scientific pieces of music of any degree of complexity or extent."

No one back then made too much of the difference between playing a chess game from start to finish and composing an original symphony, starting from the blank page. But from today's point of view, the gulf is enormous. The former can be accomplished through the linear application of rules and logic, Boole's laws of thought. The latter requires more—namely, the ability to generate new and original ideas. The former can be reduced to algorithms, while the latter (as we'll see), when we attempt to reduce it to algorithms, falls flat. Traditional computers can do the former better than any person but cannot do the latter very well at all. In that gap lies a key to the difference between analytical thought and the greater power of elastic thinking. That's right: The analytical approach we've worshipped in Western society ever since the Age of Reason is a low-level god, while the Zeus of human thought is elastic thinking. After all, logical thought can determine how to drive from your home to the grocer most efficiently, but it's elastic thought that gave us the automobile.

The Non-Algorithmic Elastic Brain

In the 1950s, many of the pioneers of information science believed that if they got the top experts together for a meeting, they could come away with a computer whose "artificial" intelligence could rival human thought. Not differentiating between analytical and elastic thinking, they saw our brains, as Lady Lovelace did, as akin to a biological version of their new instruments. They received funding for their conference, the 1956 Dartmouth Summer Research Project on Artificial Intelligence, but they did not deliver on its promise.

The most famous and influential program of that time was called the General Problem Solver, which sounds like something you'd see advertised on late-night TV, between ads for a nine-in-one blender/can opener that also cooks pasta, and knives that double as nail files. The name General Problem Solver seems grandiose, but it stemmed from naïveté about the program's potential more than arrogance.

Why not a "general problem solver"? Computers are symbol manipulators. Those symbols can be used to represent facts about the world. They can also represent rules that describe the relationships among those facts. And they can represent rules governing how all the symbols may be manipulated. In that way, the early pioneers reasoned, computers could be programmed to think. The technology of computers had changed since Boole and Babbage, but the concept hadn't.

In the naïve view, if Jane loves peach pies, and Bob bakes a peach pie, a computer can calculate Jane's love for what Bob baked—and perhaps even Jane's love for Bob—as easily as it can calculate the square root of two. But the limitations of that approach soon became apparent. The General Problem Solver was no universal genius at all. Though it could solve specific and well-defined riddles like the famous "Tower of Hanoi" puzzle, in which one attempts to reconfigure stacks of disks that slide onto vertical rods, the program choked on the ambiguity inherent in real-world problems.

To process all the novelty and change it encountered in real-world circumstances would have required both a deep understanding of the complex world and elastic thinking. But those early computers were stuck at a level somewhere between the simple scripts of the slime mold and very basic analytical reasoning.

The effort to create a computer that can execute elastic thinking hasn't progressed much since then. Today we live in a time that would have astounded Boole and Babbage, as well as those early pioneers. We build billions of microscopic Babbage-like engines onto tiny silicon chips, and perform countless of Boole's calculations every moment. But, like cancer cures and clean, cheap energy from nuclear fusion—which always seem just around the next bend—computers that can do what the General Problem Solver promised have not materialized.

In the words of Andrew Moore, who left his job as vice president at Google to run the famed school of computer science at Carnegie Mellon, even today's most sophisticated computers are only "the equivalent of really smart calculators, which solve specific problems." A computer, for example, can solve the arcane equations of physics to calculate what happens when black holes collide, but first a human must "set up" the problem by deriving the equations for that particular process from the more general theory; and no computer can create the theories themselves.

Or consider Lady Lovelace's dream: music composition. We have computers that compose complex pieces of music, and it is not at all unpleasant to listen to. There are classical pieces in the style of Mozart and Stravinsky, and jazz that sounds like Charlie Parker might have created it. There is even an app called *Bloom*, available on iTunes, that will generate, on each demand, a new and unique Brian Eno–style composition of looping instrumental passages. Eno has speculated that, with the advent of technology for such "generative music," our grandchildren might someday "look at us in wonder and say: 'You mean you used to listen to exactly the same thing over and over again?'"

Such computer music is enticing, and has its place, but it should be distinguished from new musical creations. The computer composers use lists of human-compiled "signatures"—the melodic, harmonic, and ornamental motifs created by human composers—and apply general rules to vary and intertwine them. That is mere reshuffling of old tropes, with no new ideas added. Were a human to come along and compose music that mimics Mozart or Brian Eno, or make paintings that imitate Rembrandts, we wouldn't hail their artistry—we'd call that person derivative and unoriginal.

The problem in achieving elastic thinking in computers is that though computers are heading toward ever-faster calculation, that hasn't translated to ever more *elastic* processing. And so, in the decades since those heady early days, tasks that follow explicit, easily codified rules or procedures have proved fantastically amenable to automation, while tasks that involve elastic thinking generally have not.

Consider the following paragraph:

Aoccdrnig to a rseheearcr at Cmabrigde Uinerevtisy it deosn't mttaer in waht oredr the ltteers in a wrod are, the olny iprmoatnt tihng is taht the frist and lsat ltteer be at the rghit pclae. The rset can be a tatol mses and you can sitll raed it wouthit porbelm.

There are many computer programs that can read printed text aloud, but they choke when presented with such a serious deviation from standard spelling. We humans, by contrast, have very little difficulty with it.

The surprising ease with which you can read the paragraph attests to the elasticity of your thinking. Your mind notices, without being cued, that something is not right. Then it figures out what is going on, focuses on the correct first and last letter of each word, and plays it loose with the letters in the middle. With the aid of context, it decodes the meaning with just a little slowdown in pace. The text-reading computer would attempt to match each string of letters to a word in the dictionary, and perhaps consider some common typos and spelling errors, but ultimately it would get nowhere—unless it was supplied in advance with a program tailored for that specific task.

Tasks that require elastic thinking can be exceedingly difficult to perform on a modern computer, even if they are trivial for humans. Consider pattern recognition. MIT economist David Autor talks about the challenge of visually identifying a chair. Any school-age child can do that, but how would you program a computer to do it? You could try to specify key defining features, such as a horizontal surface, a back, and legs. Unfortunately, that set of features encompasses many objects that are not chairs, such as a stove with legs and a built-in backsplash. On the other hand, there are chairs without legs that would not qualify under that definition.

A chair is difficult to define via a rational, rule-based description because the definition must embrace not just typical chairs, but a great variety of novel versions. So how does a third grader make the identification? The elastic thinking of the brain is non-algorithmic, by which I mean that we achieve our ideas and solutions without a clear definition of the steps needed to get there. (I say this regardless of whether or not the brain can be simulated by a Turing Machine, as some believe.) Instead, rather than rely on a well-thought-out and easily stated definition of a chair, the neural networks in our unconscious minds, through years of seeing examples, somehow learn to weigh complex object traits in a manner of which we are not even aware.

Some clever and forward-looking computer scientists at Google are now trying to improve on ordinary computers by finding ways to imitate our brains' neural networks. They built a machine that learned, without human supervision, to recognize the visual pattern we know as a cat. The feat required a thousand computers networked together. A child, on the other hand, can do it by age three, and while eating a banana and smearing peanut butter on the wall.

That brings us to some key differences in the architecture of brains and digital computers, which in turn tells us something important about ourselves. In contrast to our brains, computers are made of interlinked switches that can be understood through circuit and logic diagrams, and they execute their analysis by following a well-defined series of steps (a program or algorithm) in a linear fashion that is specified for the task at hand by a programmer. The Google scientists who linked a thousand such computers in a neural net performed an impressive feat, and it's a promising approach. But our brains do something vastly more impressive, forming neural nets from *billions* of cells, each connected to thousands of others. And these networks are organized into larger structures, which are in turn organized into larger structures, and so on, in a complex hierarchical scheme that scientists are only beginning to understand.

As I mentioned earlier, such biological brains can process information in a top-down manner, as a traditional computer does, or from the bottom up, which is important in elastic thinking, or in some combination of the two modes. As we'll see in chapter 4, bottom-up processing arises from the complex and relatively "unsupervised" interaction of millions of neurons and can produce wildly original insights. Top-down processing, in contrast, is directed by the brain's executive regions and produces step-by-step analytical thought.

Our executive brain is good at quashing ideas that are non sequiturs. But if we are problem-solving and happen to be plodding along in the wrong direction, non sequiturs—steps that don't follow—are exactly what we need. Sanford Perliss, a well-known defense attorney, tells of a case he heard in law school. A defendant was on trial for murdering his wife. The circumstantial evidence was strong, but the police had never found the body. When writing his closing argument, the defense attorney first tried the usual approach, summing up the evidence in an effort to persuade the jury to find reasonable doubt. But the logic wasn't working: The attorney feared he would convince no one. And then he got an idea "out of left field."

When he finally stood before the jury to make his argument, the attorney made a dramatic announcement: The supposed victim had been located. She was there, in the courthouse. He asked the jurors to turn toward the back of the room. In just a moment, he told them, she would walk through the doors, proving his client's innocence. The jurors turned in anticipation. A few seconds passed, but no one walked in. The attorney then pronounced with great bravado that unfortunately they had *not* located the woman—but if the jurors turned to look, then in their hearts there was reasonable doubt, and they should vote to acquit. It was a brilliant example of a lawyer's mind abandoning the usual step-by-step approach and taking a new direction. Unfortunately for the defendant, his attorney had not clued him in on the ruse. As a result, he himself, having *no* doubt that his wife was dead, did not turn toward the back of the room. The prosecutor pointed this out in his rebuttal, and the defendant was convicted.

You don't solve riddles through a step-by-step linear approach, nor is that how J. K. Rowling invented the Harry Potter world, or how Chester Carlson thought of the idea for the Xerox machine. It's our unsupervised bottom-up thinking that provides us with the unexpected insights and new ways of looking at situations that produce that kind of accomplishment.

We'll return to the differences between top-down and bottom-up processing, and between computers and brains, in chapter 4, and we'll examine more closely the role of those differences in producing the elastic thinking that human brains can accomplish but computers can't. But first, in the next chapter, we'll ask why brains bother to think at all. Computers do their calculations because someone turns them on and clicks a mouse somewhere. What turns our brains on?

3

Why We Think

Desire and Obsession

Pat Darcy[*] was forty-one in 1994 when she noticed an odd pain in her right arm. Then she developed a minor tremor, and it became clear that this was not simply a chronic muscle ache. She was diagnosed with Parkinson's disease. Parkinson's results from neurons dying in a part of the brain that controls your body's movements. No one knows why the neurons die off, though the dead neurons show an accumulation of a certain protein. Exposure to pesticides will increase your risk, and, ironically, smoking will decrease it.

Parkinson's patients find that they might be able to will an arm or leg to move, but their body doesn't respond as they want it to. Their speech can become slurred, their balance unstable, and their limbs stiff and painful or numb—and they may begin to shake. We know of no way to bring the dead neurons back to life, or to coax the body into growing new ones.

The cells that die are "dopamine neurons"—nerve cell factories that create dopamine and then use that neurotransmitter to send their signals to other nerve cells. They are located in the brain stem, at the top of the spinal column, in a part of the primitive midbrain called the

[*] Not her real name.

substantia nigra, which is involved in selecting the physical action, such as the initiation of motion, that you take in response to a situation. The term *substantia nigra*, which is Latin, may sound intimidating. In Latin, the phrase "Employees must wash their hands" would probably sound intimidating. But though substantia nigra sounds like something you'd hear the pope utter at Easter Mass, its meaning is mundane. It means "black substance," which pretty much encapsulated everything we knew about it when it was named in 1791, and for about 150 years after that. Its dark color comes from an abundance of melanin in the very dopamine neurons that Parkinson's affects. By the time Pat Darcy felt the symptoms of her disease, the majority of those neurons had probably already wasted away.

Dopamine neurons are found in a relatively small number of brain areas, but they are abundant in the substantia nigra. To relieve Pat's symptoms, her neurologist put her on a dopamine agonist, a drug that mimics an increase in the dopamine levels in the brain. Given the poor state of our knowledge about the disease, that's about all modern medicine can do—attempt to compensate for the action of the dead neurons by helping the survivors become more effective at transmitting their signals. Darcy's symptoms improved.

For a few years, her life was better. Then Darcy began to change her lifestyle. She had always enjoyed painting, but now she began to paint compulsively. "I transformed my home into a studio, with tables and canvases everywhere," she said. She became obsessed, painting from morning till night, and often through the night, using countless brushes, sponges, and even knives and forks. No longer painting because it gave her pleasure, she now felt an irresistible *need* to paint, like an addict craving a drug. "I started painting on the walls, the furniture, even the washing machine," she said. "I would paint any surface I came across. I also had my 'expression wall' and I could not stop myself from painting and repainting this wall every night in a trance-like state."

I once knew a drug addict. She looked malnourished and prematurely aged, with sunken eyes and an expression that said she'd do anything for a fix. That Pat Darcy painted lilies on her Maytag seems to pale in comparison, but the tragedy of any addiction is that it takes over your life and can ruin it. "My uncontrollable creativity turned into something destructive," said Darcy.

Kurt Vonnegut wrote that we humans "have to constantly be jumping off cliffs and developing our wings on the way down." We like to set up challenges for ourselves and then invent ways to overcome them. Pat Darcy's sensibilities led her to the challenge of creating art, but her dopamine therapy had amplified that natural desire into an irresistible urge.

How? As discussed, dopamine in the substantia nigra is involved in the initiation of motion (which is why a lack of it affects the mobility of people with Parkinson's). But apart from that, dopamine also plays a key role in communication among a group of diverse structures that work together in a complex manner to constitute what is called the brain's reward system.

Unfortunately for Parkinson's patients, we don't yet have the technology to deliver dopamine therapy in a precise manner, to affect only specific structures. As a result, Darcy's drug didn't just boost her low-functioning substantia nigra; it supercharged all the areas reliant on dopamine, including her reward system. And that was what caused her obsession.

Our reward system is evolution's way of encouraging us to do what it takes to stay nourished and hydrated, and to make progeny. It creates our feelings of desire and pleasure and, eventually, satiety. Without our reward system, we would feel no joy from a luscious bit of chocolate, a sip of water, or an orgasm. But it also encourages us to *think*, and to act on those thoughts, in pursuit of our goals.

When my son Alexei was a sophomore in high school, I told him that if he studied just a half hour more each day, he could get As instead of Bs. He said, "Why would I want to do that?" and looked at me as if he finally understood why I needed to see a therapist. Back then, Alexei's mind reminded me of the lawn mower we had when I was growing up. If you yanked on its starter rope hard enough, it would go into action and trim a few blades of grass, but then it would sputter and die. I could yank on Alexei's rope as often as I wanted, but without the compelling motivation that can come only from within, Alexei's brain refused to think.

Getting a computer to process information is easy. You just turn it on. But the human brain's "on" switch is internal. It is your reward system that provides the motivation to initiate or continue a chain of thought. It is what turns your information processing to matters of schoolwork or shopping or reading the newspaper or solving a jigsaw puzzle. It guides your brain toward choosing which problems to reason about, and it helps define the end point that reasoning aims to reach. As one neuroscientist put it, "There is no greater joy that I have in my life than having an idea that's a good idea. At that moment it pops into my head, it is so deeply satisfying and rewarding . . . My [reward system] is probably going nuts when it happens."

Pat Darcy's reward system inspired her to engage in the elastic thought processes involved in her artistic and creative endeavors. But its enhancement due to dopamine therapy put her interest in creating art into overdrive, depriving her of the ability to stop engaging in it.

Because of the effects it had on her behavior, Darcy's doctors eventually reduced her medication. Unfortunately, her Parkinson's symptoms then worsened, so she had surgery in which a small hole was drilled into her scalp, and a tiny probe inserted. Liquid nitrogen was circulated through the probe to destroy precise parts of the brain. That this would help seems counterintuitive, because the disease is caused by the death of cells that produce dopamine. But the surgery didn't directly address the cause of the disease; it treated its symptoms, destroying tissue whose activity is normally *suppressed* by dopamine and had become hyperactive. In Darcy's case, it brought her symptoms under control, and with the reduction in medication, her artistic urge became more tranquil and structured. "It once again became a pleasure, which upsets no one," she said.

When Thought Goes Unrewarded

If your reward system motivates you to think, what would a person be like if he were unable to experience the pleasure the reward system provides? We have insight into that question thanks to an unfortunate fellow who, in the neuroscience literature, is called Patient EVR.

Raised on a farm, EVR was an excellent student who married just after high school and by age twenty-nine had risen to the position of comptroller at a well-established home-building firm. Then, at age thirty-five, he was found to have a benign brain tumor, which was surgically removed. Despite the surgery, doctors expected him to have "no major dysfunction." EVR took just three months to recover, but when he did, it soon became clear that his thinking had a major defect.

In his everyday environment, EVR couldn't make a decision. At work, for example, if given a task such as classifying documents, he might spend the whole day debating with himself the pros and cons of a scheme based on date versus one based on document length or relevance. When he went shopping, he spent an inordinate amount of time choosing from among different brands, considering in depth every detail about them. "Deciding where to dine might take hours," one of his doctors wrote. "He discussed each restaurant's seating plan, particulars of the menu, atmosphere and management. He would drive to each restaurant to see how busy it was, but even then he could not finally decide which to choose."

EVR's doctors performed a battery of tests, none of which showed anything wrong. He had an IQ in the 120 range. When administered a standard personality test called the Minnesota Multiphasic Personality Inventory, he appeared to be normal. Another test, the Standard Issue Moral Judgment Interview, showed that he had a healthy understanding of ethics, and he seemed to have no trouble grasping the nuances of social situations. He responded knowledgeably about foreign affairs, the economy, and financial matters. So what was wrong with him? Why couldn't he make a decision?

EVR's doctors believed he had no physical deficit. His "problems are not the result of organic problems or neurological dysfunction," they said. It was the kind of dismissive, defensive response you'd expect if they'd removed a wart from the tip of his nose and now he was blaming them for his sinus headaches. True, this was the 1980s, and, compared with today, both our understanding of the brain and the technology to examine it looked like something out of *The Flintstones*. Still, when a patient has something cut from his brain and emerges with a behavior problem, you tend to suspect the surgeon.

EVR's physicians insisted that the problem was his "compulsive personality style," and that his issues after surgery reflected nothing but "adjustment problems and therefore are amenable to psychotherapy." Having received no help, EVR eventually gave up on his doctors.

In hindsight, the problem in diagnosing EVR was that all the exams were focused on his capability for analytical thinking. They revealed nothing because his knowledge and logical reasoning skills were intact. His deficit would have been more apparent had they given him a test of elastic thinking—or watched him eat a brownie, or kicked him in the shin, or probed his emotions in some other manner. For when research scientists later got hold of EVR and conducted controlled experiments on him, they found that he was decidedly *not* normal.

EVR had little capacity for feelings. There are probably plenty of people who'd argue that you could say the same thing about their spouse. But not being in touch with your feelings is different from not having any. That shrug you get as an answer to "How are you feeling?" might not tell you much, but the screaming at the television when football is on speaks volumes—the man is capable of feeling something.

Today we know enough about the brain to connect the physical damage that resulted from EVR's surgery to his mental deficits. What's relevant for us here is that among the tissue his doctors removed was most of a frontal lobe structure called the orbitofrontal cortex, a part of the brain's reward system. Without it, EVR could not experience conscious pleasure. That left him with no motivation to make choices or to formulate and attempt to achieve goals. And that explains why decisions such as where to eat caused him problems: We make such decisions based on our goals, such as enjoying the food or the atmosphere, and he had no goals.

Consider the contrast in EVR's ability to complete the tests of intellect and knowledge that his medical doctors administered with his inability to make real-life decisions. His doctors tested his knowledge and comprehension of topics such as social norms, the economy, and financial matters. In those tests, the decision criteria were externally determined: He was asked to choose the *correct* answer. That requires analytical, but not elastic, thinking. The real-life situations he faced were open-ended, with no correct answers, only preferred ones. The difference is like that between answering the question "Where is Paris?" and the question "Where would you like to vacation?" To answer the latter requires you to *formulate* and *invent* the criteria that would determine your choice. That is elastic thinking.

Evolution endowed us with emotions like pleasure and fear in order that we may evaluate the positive or negative implications of circumstances and events. Lacking any emotional reward to drive his choices, EVR's everyday decision-making was paralyzed. What's more, with no reward value attached to even completing the process of coming to a decision, EVR had no motivation to stop analyzing the pros and cons of the various options. And so, though he could pick out the correct answer on a factual exam, when faced with a real-world choice, he got stuck in an endless loop. Sadly, EVR was unable to maintain a productive work life, and was eventually fired. He then made some bad business moves and went bankrupt. Finally, his wife divorced him, and he moved back in with his parents.

We are adept at confronting novelty and change because, when faced with an unfamiliar obstacle to achieving our goals, our emotion-based reward system guides us toward elastic thinking, stimulating us to generate alternative ideas and invent a way to choose from among them. When that system doesn't function, we cannot make choices. The lesson of EVR is that emotions, especially pleasure, do not just make our lives rich—they are an integral ingredient in our ability to face the challenges of our environment. Perhaps the elusive key to success in artificial intelligence is to learn to build a computer that solves problems because it *enjoys* solving them.

Choice Overload

EVR provides a cautionary tale for us all. For even if we don't have the organic decision-making problem EVR experienced, we may still find ourselves drained by repeated demands on our elastic thinking as we make decisions, each rooted in emotion, in today's choice-rich environment. Research suggests that, when faced with too many choices or too many decisions to make, we experience a "choice overload," analogous to the "information overload" so famous in our current age. Both types of overload stimulate the primitive parts of your brain that respond to fear in life-and-death situations, depleting your mental resources, causing stress, and undermining your self-control.

William James expressed the danger of too much choice more than a hundred years ago, writing, "There is no more miserable human being than one . . . for whom the lighting of every cigar, the drinking of every cup, the time of rising and going to bed every day, and the beginning of every bit of work, are subjects of express volitional deliberation." Unfortunately, in society today, we are faced with an unprecedented torrent of choices. As Swarthmore psychologist Barry Schwartz documented, even a trip to the grocery store can be overwhelming. For example, at his own local, medium-size store, he reported finding 85 varieties and brands of crackers, 285 varieties of cookies, 61 suntan lotions, 150 lipsticks, 175 salad dressings, and 20 different types of Goldfish crackers. Yes, in just a few thousand years, we've evolved from people who'd be happy eating an undercooked beaver to individuals who obsess over whether their cracker should be Original Cheddar or Queso Fiesta.

Luckily, there is a remedy for choice overload. One can employ a decision-making strategy in which one accepts the first satisfactory option, rather than continuing to look for a superior one. Psychologists call those who do the former "satisficers," as opposed to "maximizers," who always try to choose the best. The term comes from a combination of the words *satisfy* and *suffice*. It was coined by Nobel Prize–winning economist Herbert Simon in 1956 to explain the behavior of decision-makers who don't have enough information or computational power to make the optimal choice and, rather than struggle to remedy the limitations, decide to save time and effort by making a choice despite them. But it's a concept that is just as powerful in psychology as it is in economics.

When you are choosing a video, television program, or film, do you shop around, scanning many options, instead of settling quickly on something to watch? When you're shopping for clothes, do you search forever, experiencing difficulty in finding garments you really love? When you're researching appliances, do you scour *Consumer Reports*, Amazon.com reviews, and numerous other websites to gather a mountain of information before you buy? If so, psychologists would say you tend to maximize.

We all want to make good choices, but research shows that making exhaustive analyses, paradoxically, doesn't lead to more satisfaction. It tends to lead instead to regret and second-guessing. Letting go of the idea that a choice must be optimal, on the other hand, preserves mental energy and allows you to feel better if you later learn that a better choice existed. What works when choosing shoes or a new car or a vacation plan may not suffice when choosing a doctor or a partner for what you hope will be a lifetime relationship. But for most situations, those who accept options that are good enough, rather than feeling compelled to find the optimal one, tend to be more satisfied with their choices and, in general, happier and less stressed individuals.

How Good Feelings Happen

That we have a reward center in our brains was discovered by Peter Milner, a postdoctoral fellow at McGill University who was studying the regulation of sleep. The reward system and sleep regulation may seem unrelated—and they are. But research often takes you in unexpected directions, especially early in your career. It's as if you sign up as a checker at Walmart, but the real job turns out to be shampooing dogs. That's what happened to Milner.

It's hard to imagine now, but at one time a dominant theory held that our actions could be explained solely as punishment avoidance.

That was the situation in 1954, when Milner was implanting electrodes in rat brains, targeting a structure near where the base of the brain tapers to form the brain stem. The electrodes were connected by long, flexible wires to an electrical stimulator, allowing activation of the brain region in which they rested.

One day, Milner's supervisor, a renowned psychologist named Donald Hebb, introduced a new postdoc named James Olds. Olds was still wet behind the ears, so Hebb asked Milner to show him the ropes. Soon the new postdoc was inserting electrodes himself. The experiment involved placing the rodent in a large box, with corners labeled A, B, C, and D. Whenever the animal went to corner A, the protocol called for Olds to press a button that would give a mild jolt to the rat's brain.

Olds was surprised to observe that after a few shocks from the electrode, the rat habitually returned to corner A. He also noted that if he began to stimulate the rat's brain when it was in corner B, the rat would go there instead.

The intent of the study was to stimulate a part of the brain that is involved in sleep versus wakefulness, but instead they appeared to have created a robot rat. It didn't seem like an advance that would get your face on a postage stamp, but Olds and Milner were curious about it. Milner tried to replicate the experiment with other rats, but he couldn't.

What was going on? The researchers took the rat to a nearby lab that had an X-ray machine and convinced the operator to X-ray the rat's head. That was when they saw that Olds had missed his target. He had inserted the electrode into a then obscure structure situated deep in the brain, called the "nucleus accumbens septi," or, a little more simply, the nucleus accumbens. Like *substantia nigra*, it is a grand term for a mundane message; it means "nucleus adjacent to the septum."

Olds and Milner procured other rats and began inserting their electrodes there. They also built a lever into the box so that the rats could stimulate the electrodes themselves. That was when things got really strange. Once the animals experienced the electric stimulation of their nucleus accumbens, they would continue to stand over the lever and press it incessantly, some of them as frequently as a hundred times per minute.

Like Pat Darcy so many years later, the rats had become obsessed. Male rats would ignore female rats in heat, and female rats would abandon their newborn nursing pups, just so they could continue to press the lever. Mesmerized, the rodents ceased all other activities, even eating and drinking. They had to be unhooked from the electrodes to prevent their death from starvation or thirst.

Today we know why. Under ordinary circumstances, the achievement of a goal comes through effort that we expend over time. As a result, your reward system evolved not just to provide pleasure when you reach a goal, but to continually predict the consequences of what you are doing and reward you at every stage.

When you are hungry, you don't just feel satisfied at the end of your lasagna; you enjoy each bite along the way. When you drink your wine, you enjoy each sip. And when you think about an issue, if you seem to be headed in the right direction, your brain likewise provides you with ongoing feedback to encourage you to continue—subtle positive feelings of progress, confidence, or impending accomplishment.

As a goal is achieved, your body generates feedback to diminish the reward value of continuing the activity. The pleasure you felt at the beginning fades, and before long you'd just as soon be watching *I Love Lucy* reruns. This causes you to stop the behavior, rather than engaging in it without end. That's what happens when you eat—as your body senses that you've ingested enough food, further bites bring diminished brain activity. Similar pleasure response and satiety feedback occurs for other pleasures, such as sex.

The nucleus accumbens that Olds had inadvertently stimulated is a reward system structure involved in that process, in particular with regard to a basic need, such as obtaining food, water, or sexual contact.

The signal for the nucleus accumbens to spring into action comes from another reward system structure, called the ventral tegmental area (VTA). The interaction between those two structures can be complex and involve other structures, such as the prefrontal cortex, but in simple terms, satiety occurs when the body, sensing that we've had enough, communicates that to the VTA, which diminishes or halts its signal to the nucleus accumbens. If we are thirsty and drink water, for example, the VTA signals the nucleus accumbens and we experience pleasure, but with each successive sip, the signal diminishes, and we eventually lose our motivation to continue drinking.

By pressing the lever, the rats were stimulating their nucleus accumbens *directly*, which overrode the role of the VTA. To the rats, each press must have felt like a swallow of water at a moment of thirst, a mouthful of food that quenches hunger, or perhaps even an orgasm, with no diminishment due to repetition. Desire and reward, without satiety, is like a car with the pedal to the metal and no brakes. That is what happened, in effect, to Pat Darcy when she flooded her brain with the dopamine agonist.

Patient EVR was insensitive to the reward value of his thoughts and actions; Pat Darcy was a slave to them. Healthy individuals fall somewhere in between. How "reward dependent" are you? Psycholo-

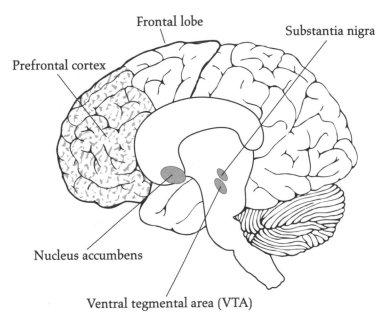

The substantia nigra, nucleus accumbens, and
ventral tegmental area (VTA) in context

gists have developed a thirteen-statement questionnaire to assess the extent to which the prospect of reward motivates a person. To test yourself, just rate each statement that follows with a number from 1 to 4, as explained below.

1 = strongly disagree
2 = disagree
3 = agree
4 = strongly agree

Here are the statements:

1. ___ When I get something I want, I feel excited and energized.
2. ___ When I want something, I usually go all out to get it.
3. ___ I will often do things for no other reason than that they might be fun.
4. ___ When I'm doing well at something, I love to keep at it.
5. ___ I go out of my way to get things I want.
6. ___ I crave excitement and new sensations.
7. ___ When good things happen to me, it affects me strongly.
8. ___ If I see a chance to get something I want, I move on it right away.
9. ___ I'm always willing to try something new if I think it will be fun.
10. ___ It would excite me to win a contest.
11. ___ When I go after something, I use a "no holds barred" approach.
12. ___ I often act on the spur of the moment.
13. ___ When I see an opportunity for something I like, I get excited right away.

Total: ___

The average result on this assessment is 41, and about two-thirds of those tested scored between 37 and 45 (the maximum is 52). If you scored low, you are less driven by rewards than the average person and may be more focused on the journey than the destination. You are also likely well equipped to find balance in your life. On the other hand, if you scored high, you probably exhibit a strong drive to initiate or increase progress toward your goals. You may even have built your life around the rewards that come from accomplishment. That's good, in that it drives you toward achievement, but it might mean that you experience particularly acute emptiness during periods of unemployment, when you retire, or (as in my case) when you are between projects. You may also, on occasion, tend toward impulsivity and risk-taking behavior. And you might tend, in your thinking and in your decisions, to be overly influenced by the promise of social approval or support, sexual pleasure, or monetary gain.

16% fall here	68% fall here	16% fall here
13 37	41 45	52
reward insensitive	average	reward sensitive

Distribution of reward sensitivity scores

An unflappable drive toward goals is one of the keys to personal and professional success—psychologists call it *grit*. It's what compels you to keep working until your work of art is just right, or to persevere when you hit an impasse, until your elastic brain generates the idea that's the key to solving your problem. But impulsivity can get us in trouble, risk-taking is a double-edged sword, and an exaggerated focus on social approval, sex, or monetary rewards can lead to unhappiness. You can rein in those tendencies once you are aware of them, and the questionnaire helps you achieve that awareness.

The Rewards of Art

Few people enjoy solving the rote "drill or kill" problems many schools assign when teaching elementary math. We call that work "mindless," because it requires no real thought, just the choice of which fixed algorithm to apply. But most of us enjoy some sort of challenge to our thinking skills—activities like card games, chess, crossword puzzles, jigsaw puzzles, riddles, Sudokus, repairing cars. Those are all, in a sense, problem-solving activities, but unlike the rote math problems, they require idea generation and other aspects of elastic thinking.

As I mentioned, elastic thinking is useful to our species, and so your brain takes pains to encourage it. As you take steps to solve a problem, the subtle stream of reward produced when you seem to be making progress is your emotional brain's way of keeping your attention trained along those lines. We've all experienced a conscious gut feeling that our thinking is on the right track, but we also experience unconscious coaxing that we are not mindful of, which guides the direction of our thinking nonetheless.

It's easy to understand why we evolved to take satisfaction from problem-solving, but as Pat Darcy's case illustrates, our brains have also evolved to make us happy when we engage in the arts. The exercise of artistic skills predates even the origin of our species. One million four hundred thousand years ago, our predecessor *Homo erectus* created the first aesthetic artifacts we know of: symmetrical hand axes. These look attractive, and that must have been the purpose of their symmetry, for to make them symmetrical employing the bone, antler, and stone manufacturing tools of the day required a great investment of time and energy, and added little to their utility. Artsy hipsters today may adorn themselves with retro rings and earrings, but if you really want to be retro, try carrying around a symmetrical hand ax.

That the exercise of elastic thinking skills such as idea generation, pattern recognition, divergent thinking, and imagination is inherently rewarding is the reason people have always put energy into the arts, despite the lack of material reward (for most). In fact, material recompense can even get in the way of the pleasure we feel in such activities. Consider, for example, how the great Russian writer Fyodor Dostoyevsky responded when a Russian publisher paid him a fairly large advance to write a novel. Note that he had *not* been given strict guidelines regarding what to write; he had merely been asked to write something engaging in exchange for money. Despite that, in a letter to a friend, Dostoyevsky wrote, "I believe you have never written to order, by the yard, and have never experienced that hellish torture." The hellish torture he referred to wasn't just the great novelist being a drama queen; the prospect of being paid for his work gave Dostoyevsky writer's block.

Dostoyevsky's was not an isolated case. Many recent studies in social psychology suggest that monetizing creative output can disrupt the processes that lead to innovation. That contradicts the ideas of traditional psychology, which is full of papers investigating the importance of reward in encouraging or even controlling a person's behavior. But offering an extrinsic reward for an *intrinsically* enjoyable behavior can be counterproductive. Difficulty in original thinking arises, says psychologist Teresa Amabile, when you "try for the wrong reasons."

Our brains reward original and artistic thinking because those skills are important to any animal's ability to respond to change and unpredictability. It makes sense, then, that many animals make a display of their artistic nature when advertising themselves to a mate. Peacocks preen, and songbirds warble. Young male zebra finches learn to sing by imitating adult males and then, upon reaching sexual maturity, churn out diverse, novel tunes of their own. Might artistic talent play a similar role in human mating? Might it, on an innate and unconscious level, indicate to a prospective mate that one has genes that could play a role in survival?

Evolutionary psychologists Martie Haselton and Geoffrey Miller tested that hypothesis by studying how women's taste in men changes at different stages of their ovulatory cycle. Haselton and Miller knew that when women judge male attractiveness while they are at peak fertility, just before ovulation, they unconsciously give more weight than usual to indicators of evolutionary advantage such as upper-body musculature, beard growth, and jaw size. Those big guys pumping iron at the gym are chick magnets, but little do they know that the strength of their attractiveness depends on a woman's time of the month. Haselton and Miller reasoned that if imagination is also a signal of mating fitness, artistic talent ought to have the same varying effect on women as a great set of pecs.

To find out, Haselton and Miller gathered forty-one women in their twenties and recorded data on the timing of their cycles. Then they presented each of them with a carefully written description of two young men. The descriptions were designed so that the men had comparable qualities, except that one was artistic but poor, while the other was average in creative intelligence but wealthy. Though each woman would weigh those traits according to personal taste, the question was whether a given woman, at a time of month when her body was ready for reproduction, would have a greater tendency to choose the man more likely to generate creative offspring. If so, that would support the idea that artistic ability is a way we signal reproductive fitness.

After reading the vignettes, the women rated each man's desirability on a scale from 1 to 9, and responded to the written question "Whom do you think you might find more desirable for a short-term sexual affair?"

The results were enlightening, especially if you are a poor male artist. Women's desirability ratings of the creative but poor men were strongly correlated with their degree of fertility, while fertility had no effect on their rating of the non-imaginative but rich men. As to the question of whom they would choose for an affair, the effect of fertility timing is quite striking. When their fertility was high, 92 percent of the women chose artistic ability over wealth, but when it was low, only 55 percent did so. It's a cliché that artistic types have no trouble connecting with members of the opposite sex, but it's nice to know that at the root of that connection is the evolutionary importance of imagination.

Attention Deficit, Elasticity Surplus

In the early 1990s, a young University of Georgia pioneer in educational psychology named Bonnie Cramond noticed an oddity in the slowly growing literature on attention-deficit/hyperactivity disorder (ADHD). The children described in that research seemed to share many traits with those described in the research on gifted children. For example, both children with ADHD and gifted children were branded as distractible and having a gluttonous appetite for activity.

Those sound like negative traits. In fact, when the disorder was first described, in the early twentieth century, doctors thought it was attributable to some sort of mild brain damage. That idea had been abandoned by the 1990s, but there was still a considerable stigma associated with an ADHD diagnosis. That bothered Cramond. What's more, she suspected that ADHD could actually be *beneficial* to one's thinking. Could those characteristics of ADHD be related to positive qualities—to ambition, productivity, and the ability to rapidly generate ideas?

Cramond decided to administer what was essentially a test of elastic thinking to children diagnosed with ADHD and, conversely, to administer a test for ADHD to a group of children in a "scholars' program." She found astonishing overlap. A third of the ADHD group scored high enough to qualify for the elite and super-selective scholars' program, while a quarter of those in the scholars' program were diagnosed with ADHD—four to five times the prevalence in the general population. For Cramond, that work was the beginning of a long career studying gifted children.

Today, ADHD carries little stigma, and kids are sometimes misdiagnosed with the condition to satisfy parents seeking a "cure" for what is simply their child's natural and healthy, highly active state. The misdiagnosis issue is ironic, because in recent years we've made great progress in understanding ADHD. We can now explain Cramond's results at the neural level, which brings us back to the reward system and its role in what motivates people to explore both new ideas and new places.

There is no single brain structure or system responsible for all the traits of ADHD. But the most critical traits can be attributed to the same ventral tegmental area–to–nucleus accumbens circuit in the reward system that Olds had stumbled upon. In ADHD, the dopamine receptors in those structures are impaired, resulting in a weakening of the brain's reward pathway. As a result, the steady stream of feel-good reinforcement that keeps others moving toward their goal is diminished.

One consequence is that those with ADHD have difficulty performing some of the routine tasks of everyday life. But the disorder can also cause the opposite effect. Since it can make everyday life feel routine and boring, the brain tries to compensate by seeking more stimulation. As a result, when an ADHD brain comes upon a task it finds truly interesting—that is, a task that briskly stimulates the reward circuits—it obsesses over it and becomes hyperfocused.

The most famous ADHD trait is the one that concerned Cramond. It occurs when the weakened stream of reward is not sufficient to prevent one's attention from flitting from an issue at hand to stimuli in the environment, or to thoughts produced elsewhere within one's own brain. As a result, like the children in a classroom overseen by a lenient teacher, the neural circuits of a person with ADHD shout out ideas, with little focus or censorship.

The intruding thoughts can send the individual off track, leading to a shift from one goal to another before that goal is achieved. But the stray thoughts sometimes prove relevant. When they are, they can produce unusual but constructive connections and associations that wouldn't have been thought of by "normal" people, making individuals with ADHD better at skills like idea generation and divergent thinking. For better *and* for worse, the thinking of those with ADHD is less constrained and more elastic. So, although many see ADHD as a disorder, one could also view it as tailor-made for today's turbulent and changing environment. ADHD might be, as Cramond speculated, an advantage at this stage of our evolution.

That point of view is supported by an interesting new theory, which states that we evolved ADHD as an adaptation to the demands of change when we lived as nomadic hunter-gatherers. Those nomads lived in an environment that in some ways was much like our civilization today: ever shifting, and full of unpredictable threats. In that context, elastic thinking, flexible attention, and a thirst for adventure, especially exploration, could have been beneficial.

The theory was tested in a study of the Ariaal, a Kenyan nomadic tribe. They had always been wandering animal herders, with low body fat and chronic undernutrition. Then, about forty years ago, some of the nomads split off from the main group and settled in one place to practice agriculture. Recently, an anthropologist from the University of Washington studied, in both groups, the frequency of a variant of a gene that has been linked to ADHD. He found that among the wanderers, who confront constant change, those with the ADHD gene were, on the whole, better nourished. But among those who had settled, those with the trait were significantly undernourished.

The nomadic Ariaal with ADHD apparently were better equipped to thrive in their tempestuous environment, while the settled Ariaal with ADHD were at a disadvantage in the many agricultural activities that require sustained focus. There is a lesson for us in that. Until just a couple of decades ago, our society mirrored the settled Ariaal, and ADHD might have been a hindrance. But in today's turbulent times, we are better likened to the nomadic Ariaal, so ADHD may be an asset.

ADHD is generally a condition of the immature brain, and when children grow up, they usually grow out of it. But whether or not we have ADHD, all of us have a greater or lesser propensity to explore or exploit, to wander or focus. Occupational research theorist Michael Kirton was ahead of his time when, in the 1970s, he captured that same distinction in his "adaptors and innovators" theory of cognitive styles.

Kirton described "adaptors" as focused but rigid individuals who "prefer to do things better by marshaling tried and true methods." They tend to be prudent and cautious, and seem impervious to boredom. They can seem "stuffy and unenterprising, wedded to systems and rules," wrote Kirton. "Innovators," on the other hand, are elastic thinkers who like to find new approaches to problems. They are often distractible and poor at time management, and they produce less familiar, and sometimes less acceptable, solutions that, in the corporate world, often encounter resistance. They can also appear abrasive, even to one another, Kirton wrote.

We may each have a preference for one type of thinker or the other, but companies need both, and those that don't have the proper balance, Kirton argued, run into trouble. That can be true of our personal relationships as well: An individual on one end of Kirton's spectrum is often best paired with an individual on the other end. By embracing the differences, the rule follower and rule breaker can balance each other and create a pair with the benefits, but not necessarily the drawbacks, of each individual's personality.

The Pleasure of Finding Things Out

Imagine the following scenario, occurring several million years ago: A primitive human of the species *Homo habilis*, the predecessor of *Homo erectus*, is wrestling down some small creature when he rolls over a sharp rock and slits open his own skin. After overpowering his prey, he is about to tear into its tough flesh with his teeth when his mind makes an association: *The sharp rock tore my skin; I want to tear this animal's skin; I can use the sharp rock.* For the millions of years of *Homo habilis*'s existence, that crude stone cutter was its only original creation.

Now flash forward a million and a half years. It is the early 1990s, and Jerry Hirshberg, president of Nissan's California design studio, is wrestling with the design for the new Nissan Quest. One day, while driving down the road, he spots a couple at the curb struggling to get the back seat of a competitor's minivan out of the way so they can slide in a couch. A thought immediately pops into Hirshberg's mind: Install tracks that enable the driver to fold up the back seats and slide them forward to make space. Thus was born one of the most popular features of the Nissan Quest's design.

Both inventions were the result of someone's brain forming an association among seemingly unrelated ideas. Different eras, different species, same route to discovery. In nature, different atoms collide and combine to create molecules with properties unlike those of the atoms that form them. In our minds, one neural network in our brains overlaps with and activates another, and we compound diverse concepts and observations to form new ones. Though original thought in the arts, science, business, and personal life differs in its goals and context, at the level of neural networks, all those modes of thought arise from the association of different concepts in your brain.

The mental equipment we use to solve business problems—or adjust to changing conditions in our personal lives—is the same as the equipment we use to explore or create new art and music, or theories in science. Just as important, the thought processes we use to create what are hailed as great masterpieces of art and science are not fundamentally different from those we use to create our failures.

One aspect of those thought processes is how intrinsically rewarding they are, due to the pleasure signals that our brains experience as we generate ideas. That is why there is wisdom in the saying that it is the journey that counts, and not the end point. In fact, we often don't know how that end point will be valued by society until long after our act of creation. Think of Vincent van Gogh, who sold very few paintings in his lifetime. Or Copernicus's heliocentric picture of the solar system, which impressed no one until Galileo adopted it some seven decades later. And then there is Chester Carlson, who, as I've mentioned, invented the photocopier. That was in 1938, but he couldn't sell it, because companies—including IBM and General Electric—considered it a crackpot idea. Why would anyone want a complicated copier when people could just use carbon paper?

I was lucky enough to learn many years ago to appreciate even the thought processes that lead to "small ideas," or to failed attempts. I learned it from the great physicist and Nobel laureate Richard Feynman. Feynman brushed against a "sharp rock" of an idea when, as a graduate student in the 1940s, he came across an observation made by Paul Dirac, one of the fathers of quantum theory. He made a mental association between Dirac's comment and some ideas he was already thinking about. After years of hard work, that led him to invent an entirely new—and exotic—way to look at quantum theory, and a new mathematical formalism, called Feynman diagrams, to go with it.

Like the *Homo habilis* stone tool, Feynman diagrams, in physics, are everywhere, the basis of much of the fundamental work in that science today. But if Feynman's plan had failed—if his mathematics had been proved, in the end, to have a small flaw—his ideas would have been no less imaginative. In fact, Feynman sometimes took great pleasure in describing to me original ideas he'd invented and worked on that had led nowhere. Though scientists rightfully treasure only theories that work, we can also recognize the degree of intellectual beauty in a proposed theory, whether or not it ultimately proves successful.

As for Feynman, he didn't see his most famous breakthrough as being different from any of the other problems he solved in his long career, whether big or small, or the little puzzles of ordinary life that he had to face like the rest of us. That a person who made a monumental and revolutionary contribution to his field could find equal pleasure in solving problems of far less significance is a testament to the fact that the exercise of elastic thinking is intrinsically rewarding. Feynman made that point especially well in a 1966 letter to one of his former Ph.D. students who, years after graduating, had written him and apologized for doing work that was not sufficiently important. To that, Feynman responded:

Dear Koichi,

I was very happy to hear from you, and that you have such a position in the Research Laboratories.

Unfortunately your letter made me unhappy for you seem to be truly sad. It seems that the influence of your teacher has been to give you a false idea of what are worthwhile problems . . . A problem is grand in science if it lies before us unsolved and we see some way for us to make some headway into it. I would advise you to take even simpler, or as you say, humbler, problems . . .

You met me at the peak of my career when I seemed to you to be concerned with problems close to the gods. But at the same time I had another Ph.D. Student (Albert Hibbs*) whose thesis was on [the mundane topic of how] the winds build up waves blowing over water in the sea. I accepted him as a student because he came to me with the problem he wanted to solve . . .

No problem is too small or too trivial if we can really do something about it.

You say you are a nameless man. You are not to your wife and to your child. You will not long remain so to your immediate colleagues if you can answer their simple questions when they come into your office. You are not nameless to me. Do not remain nameless to yourself—it is too sad a way to be. Know your place in the world and evaluate yourself fairly, not in terms of your naïve ideals of your own youth, nor in terms of what you erroneously imagine your teacher's ideals are.

Best of luck and happiness.

Sincerely,
Richard P. Feynman

---

* Hibbs, who died in 2003, went on to become a noted scientist at the Jet Propulsion Laboratory, in Pasadena, California.

4

The World Inside Your Brain

How Brains Represent the World

In order for any word or concept to be the object of your thoughts, that word or concept must be represented in a neural network within your brain. Aristotle made an analogous point more than two thousand years ago. Though he knew nothing of neural networks, he argued that human thought is based on internal representations of the world, and he distinguished between the image that reaches our eyes and the indirect perception of it in our thoughts. He also believed that women were deformed males and had fewer teeth. He was wrong about many things. But on this, regarding human thought, his was an accurate and important realization. For if there is no video screen in our brains that represents directly the optical data our retinas capture, then our brains must be translating and encoding that data, and any such process can both bias and restrain our thinking.

It's not just the data from our senses that must be encoded in our brains; it is all data, from the fact that ice is frozen water to the conclusion that Satan is not a good name for your dog. And, as is true of the encoding of sensory data, the manner in which knowledge, ideas, and other information is represented has a substantial effect on the way you think about that information.

For example, suppose you memorize a phone number. One obvious way to store that information is as a string of numerals, or as an image, as if you had written it down. But your brain does not store it that way. To see that, try reciting a phone number backwards. It's easy to do if the phone number is written on a piece of paper, but difficult if you are reading it from your memory. That limitation arises from the manner in which your brain *represents* phone numbers.

The issue of how to represent that which is to be "thought about" is an issue that must be solved by all information-processing systems. For example, in 1997, IBM made headlines with a computer named Deep Blue, which defeated the reigning human chess champion, Garry Kasparov, in a six-game tournament. The first task the IBM team faced in designing Deep Blue was to determine how the program would represent the game internally, in the guts of the machine. They decided to create a tree of possible moves and responses and a set of rules for rating the desirability (or "value") of any given arrangement of pieces on the board. That representation determined what Deep Blue did when it "pondered" a move: It analyzed its tree of board positions.

Kasparov's brain represented the problem of playing chess not as a tree of moves, but in a much more powerful way—as a collection of meaningful patterns. He viewed, as a single unit, small groups of pieces mutually protecting one another, together attacking a piece, or controlling certain squares. Neuroscientists estimate that he could recognize about 100,000 distinct board positions composed of such clusters of pieces.

To represent the game in terms of patterns of pieces is natural for the bottom-up neural networks that make up the human brain, while the tree-of-possibilities method the IBM programmers used is a natural way to represent data in a traditional top-down computer. The brain's approach is tailored to elastic thinking. It facilitates analysis in terms of overall strategy and principles, and the ability to improve by learning. The tree search approach is tailored to the step-by-step logical analysis that computers apply. It reduces each move decision to one huge mathematical calculation, which yields an answer but no conceptual understanding, and far less potential for learning.

Given unlimited time, a tree search could, in principle, always produce the optimal move. Since, in practice, the time allotted is not unlimited, the quality of the computer's choices is a reflection of its hardware speed. Deep Blue was far faster at evaluating chess positions than was Kasparov—it could evaluate a billion in the time it took Kasparov to assess just one. That Kasparov gave Deep Blue a run for its money despite that speed disadvantage is a testament to the potency of the elastic thinking that a human brain can employ to set up and analyze problems.

In the past decade, programmers have designed machines that play other kinds of games, with equally impressive results. In 2011, for example, IBM's computer Watson, with four terabytes of disk storage and access to 200 million pages of content, beat the reigning human champions at *Jeopardy!* Meanwhile, processors have become so fast that a $100 chess computer can now easily dispose of human world champions.

In recent years, the computing community has realized the superiority of the manner in which biological systems process information. As I mentioned earlier, they are now trying to copy it, designing software that mimics our brains' bottom-up neural networks. The effort has set off what has been called an "arms race" for talent in artificial intelligence. Google—after getting its feet wet creating the program that learned to recognize cats—led the way in adopting the nontraditional approach (see chapter 2). Facebook, Apple, Microsoft, and Amazon have now also jumped on the bandwagon.

The effort has already borne fruit—for instance, in a computer that can beat the best human at the game Go and a greatly improved version of Google Translate. Though an improvement on the traditional approach, the neural net systems of today have internal representation tailored specifically to the application they will be used for, and have no ability to adapt their processing if the task they were designed for is altered—much less, to apply their intelligence in a wide range of diverse domains. They are excellent at learning in "a highly structured situation," said one AI expert, but still, "it's not really human-level understanding."

As a result, even today's most advanced computers are nothing like the General Problem Solver hoped for in the early days of artificial intelligence.

Computer scientists had to build one machine to play Go and another to power the process of language translation. A single human brain can handle both tasks, and it can do so while controlling your balance so you can stand on one foot. Such flexibility is obviously necessary in animal brains, for we face a multitude of situations in life and cannot have a separate brain for each. To solve the unforeseen problems faced by complex life-forms, we animals have evolved an elastic mind that can create representations spontaneously, and without outside intervention—the skills necessary for survival in our world of flux. That's the miracle of biological information processing. And so, if you want to make a General Problem Solver today, the best way is still to find a mate and create a new human being.

How Brains Create Meaning

Think of what happens when an event as simple as the activation of a doorbell occurs. It is easy to make the mistake of believing that our perception of that event as a ringing sound is the physical reality, but it isn't. The physical reality of the doorbell's action is the wavelike propagation of a disturbance of air molecules.

A microphone that picks up the doorbell's sound would represent it as the modulation of an electrical current that could be transmitted, say, to a speaker that could read it and reproduce it. A radio transmitter would represent the same physical phenomenon as the modulation of an electromagnetic wave. A computer would represent it by a series of zeros and ones encoded in the quantum state of its circuitry. A snake resting in your home would sense the doorbell's activation from the manner in which the air vibrations jiggle the floor upon which its jaw rests, and create from that sensation its conception of events—whatever that is.

In our own brains, the doorbell's physical sound is transmitted by the ear and represented by the state of a network of neurons in the auditory cortex in our temporal lobes. We experience it as a ringing.

But that representation is no truer than the other four I just mentioned. It is merely a fabrication that allows us to process the information and calculate an appropriate response.

Some people, with a condition called synesthesia, will perceive the doorbell's activity as a color as well as a sound. That some human brains translate the vibration of air molecules into the perception of a hue may seem odd. From the point of view of physics, however, the representation of the bell as the sensation we call sound is no more natural than perceiving it as the sensation we call color. In fact, no one knows what a snake or a bat or a bee *experiences* when it perceives a doorbell—or how an intelligent alien would experience it—for there is no reason to believe they do so via the same ringing sensation that we experience.

In whatever manner an organism represents the physical sound, that is only the beginning. All species, if they are to survive, must process and react to important stimuli in their environment, and so, to their sensory input, they must assign meaning.

One of the key features that distinguish mammals is that their brains assign many levels of meaning, in a manner that is more sophisticated than in any other type of animal. We experience a doorbell as a ringing sound, but it also has associations that may signify interruption (that solicitor again) or social connection (my friend is here) or gratification (the FedEx man is bringing the cashmere sweater I ordered). That single disturbance of air molecules triggers a cascade of related meanings—physical, social, and emotional. So although in school we learn that the defining traits of mammals are that they have hair, give birth to live young, and suckle their babies, just as important is the unique way that mammals *think*.

One of the mammal brain's tricks in creating meaning is to group diverse elements into a single compound unit, and to group compound units into still higher-level units, and so on. Scientists call the ideas and groups of ideas represented by those hierarchies, fittingly, *concepts*. For example, the concept "Grandma" may include traits such as smile lines, gray hair, and "keeps her teeth in a jar." Whatever it includes for you, the concept is also subsumed under a larger concept, "grandmothers," which is itself a subset of the concept "older people."

Imagine spotting your grandmother unexpectedly. How do you process the data your eye picks up? The visual data of her skin coloring, eyes, hair, etc. is quickly transmitted to an area of your brain called the visual cortex, but it takes some milliseconds before you give that data meaning. If she is wearing gold sunglasses and a hat decorated with plastic bananas and pears, and if you spot her out of context while you are on vacation in Hawaii—where you weren't expecting her to be—her identity might register only after several seconds and feel like a minor epiphany. The delay is indicative of the processing that is occurring in your brain. But what kind of computation is that?

We don't yet fully understand the process, but we do know that your brain is not literally registering each bit of the woman you see as optical data, the way a computer might, pixel by pixel, then searching a database of images, and finally matching the data with a stored image of Grandma. That would be exceedingly laborious, because sometimes you see Grandma in bright light and sometimes in deep shadows, head-on or from the side or the back, wearing a huge hat with fruit on it or hatless, laughing or frowning, etc.—the variables are practically infinite. If our brains searched a database of Grandma shots, we'd have to store all such images to represent her, or else we'd need an algorithm to generate them from some standard views. An incredibly fast computer like Deep Blue can get away with that approach to information processing, but our human brains can't.

Instead, a higher level of processing is going on, related to clusters of pixels: her features. Just as, to Kasparov, it was clusters of pieces that formed the meaningful units, to you it is a collection of features (including nonvisual traits such as personality) that form your representation of your grandmother, your Grandma concept. We know this because there are neurons in your brain that will fire whenever you see her, but those same neurons fire if you simply see her name written out in text, or hear it spoken, or are reminded of some aspect of her.

Neuroscientists call the neurons in networks that represent concepts "concept cells" or "concept neurons." We have networks of concept cells for people, places, things—even for ideas such as winning and losing. I used the image of your grandmother to illustrate the idea of concept cells because they used to be called *grandmother cells*. The term was invented when neuroscientists didn't think that such cells existed, and it was meant to be derisive and sarcastic, as in "You can't really believe that your brain reserves a network of cells for thoughts of your grandmother!" But scientists changed their tune when those cells were actually discovered in 2005, and they also changed their terminology.

In those first experiments, scientists took advantage of electrodes that had been implanted deep into patients' brains in the course of treatment for severe epilepsy. The electrodes allowed them to observe the individual neurons in their subjects' brains responding to photos of places like the Eiffel Tower and the Sydney Opera House, and famous people like the actresses Jennifer Aniston and Halle Berry. The researchers were shocked to find that the same network could, for example, pick out Berry as viewed from different angles, and even when masked as Catwoman. Today researchers believe that humans have, in this regard, by far the greatest capability of any mammal. We are capable of encoding in our neurons tens of thousands of different concepts, each composed of a network of about a million concept neurons—about as many as are in an entire wasp brain.

Concept networks are the building blocks of our thought processes. Each of those networks can be accessed independently. The fact that neurons are shared among different networks seems to be the root of the associations we make between different concepts, for it allows the activation of one neural network to spread to another. When we are faced with a question or come upon new information, we operate on those concepts, perhaps merging or splitting them or calling up a new concept on the basis of an association. By stringing such thoughts together, we are led to conclusions. Every concept we ever conceive takes the physical form of a constellation of the neurons in a concept network. They are the realization, in hardware, of our ideas.

Ours is a far more complex process than occurs in a computer, an insect brain, or even the brains of other mammals. It allows us to face the world armed with a capability for an astonishing breadth of conceptual analysis. That's why, now that most of the existential struggles of the wild are behind us, we humans can turn our powers to pursuits not seen in the natural world. We can create Velcro, quantum theory, abstract art, and bacon maple doughnuts, because the elasticity of our thinking allows us to move beyond the existing world of our senses and invent *new* concepts. So while other animals have to chase down their prey across vast fields, we get to run on a treadmill, then stick a box of Lean Cuisine in the microwave, and feast on a concoction of autolyzed yeast, maltodextrin, sodium aluminum phosphate, and seventy other ingredients that the manufacturer calls "sesame chicken."

The Bottom-up Brilliance of Ants

Once information is represented in the brain, what happens next? How do brains process that information? The neurons in our brains are, in one sense, simple objects. Each receives thousands of electrochemical signals per second from the other neurons to which it is connected. Like the zeros and ones that constitute the language of digital computers, these signals come in two types: excitatory and inhibitory. The neuron applies no intelligence when assessing those input signals; it simply adds up the excitatory signals and subtracts from them the inhibitory signals. If the net input over a short period is large enough, the neuron fires, sending its own signal (which may be either excitatory or inhibitory) to the other neurons to which it is connected. How do the thoughts and intellect of all animals arise from this primitive decision-making prowess of individual neurons—simply whether or not to fire?

If the behavior of a mother goose is a good model for automatic, nonthinking behavior, the insect world provides a potent example of how intelligent information processing can arise from simple rules obeyed by a large number of individual components. That's because, faced with a challenging and often changing environment that overwhelms their simple preprogrammed abilities, certain insects evolved a method of inventive group processing that, like our neurons, creates an intelligent response from a group of unintelligent components, the individual insects.

The insects that do this, which include ants, bees, wasps, and termites, are called social insects. From the point of view of evolution, they are the most successful of all insects. Though they make up just 2 percent of the insect species in the world, they thrive in such great numbers that they form more than half the earth's insect biomass. In fact, though each individual is less than one one-millionth the size of a human, if all the ants in the world could be placed on a scale, their weight, taken together, would equal that of all the humans in the world.

The term *social insect* is in a way a misnomer, for these animals don't care one bit about their cohorts. They don't have friends, and if they hang around at cafés, it is to eat the morsels you drop, not to meet with their friends. In fact, that's my point: The members of social insect species are mindless automatons, each responding to its environment through a set of simple programmed scripts. But what distinguishes the social insects is that, over millions of years of evolution, those mindless scripts have developed in a way that, taken together, allows them to process information in a new way. As individuals, their mode of information processing is scripted and rigid, but as a group it is elastic. So as a group, though not as individuals, they can evaluate complex new situations and take meaningful action. They have a collective intelligence that is called, in the terms of mathematical complexity theory, an "emergent phenomenon."

To see how that works, consider how ants adjust the way they explore for food when the physical boundaries of their available territory shrink or expand. Since there is no ant in charge, there is no central plan. Yet if you place the ants in a ten-by-ten-foot arena and then suddenly double each dimension, the ants will process that information and change their pattern of exploration to effectively explore the larger area. Though no single ant comprehends what has changed, as a group they recognize the change and respond to it. What looks like intelligent behavior on the group level is but a simple algorithm on the level of the individual ant: Each ant, with its antennae, senses when it encounters another ant and, employing a fixed formula, adjusts its exploration path according to the frequency of those encounters.

That's a simple example, but the same kind of unsupervised reasoning allows ants, as a collective, to accomplish many intelligent feats. Army ants organize hunting raids involving up to 200,000 workers. Weaver ant workers create chains made from their own bodies, allowing them to cross wide gaps and tug leaf edges together to form a nest. Leaf-cutter ants chop the leaves off plants in order to grow fungi. Arizona harvester ants send foragers looking for food, but if it rains and there is damage to the nest, those ants change jobs to provide more maintenance labor to clean up the mess. All of this is achieved without any "executive" ant orchestrating the group's attention, reasoning, planning, or actions.

On the whole, colonies of social insects exhibit so cohesive a collective mind that some scientists like to think of the colony, rather than the individual ants, as the organism. That applies even to their reproduction, says Stanford scientist Deborah Gordon. "Ants never make more ants; colonies make more colonies," she says.

The process goes like this: Each year, on the same day—no one knows how the colony accomplishes that feat of timing—each colony sends out its winged males and virgin queens, who fly to a mating ground, where they copulate. Then all the males die, while each queen flies to some new spot. There, she sheds her wings, digs a hole, lays eggs, and starts a new colony. With that, the original colony has reproduced. That colony, with its queen, will live for fifteen to twenty years. Each year, she will lay more eggs to replenish it, still using sperm from that original mating (Most of her offspring are wingless workers incapable of reproduction, but some are new queen ants and the males who exist solely to fertilize them).

If you think about it, the way social insect societies function is completely foreign to us. Our corporations and organizations have hierarchical structures, with an individual or small group at the top directing the activities of those below, who may in turn govern the activities of those at lower levels. To have a country or a company with no one in charge is virtually inconceivable for us. We call it anarchy. But the ant queen, unlike human royalty, does not hold authority or implore other ants to carry out any action. No executive ant directs the behavior of any other ant. That is how ant colonies operate—as many as half a million ants do just fine with no management at all.

The evolutionary goal of all organisms is to understand and react to their environment in a manner that is effective enough that they survive to reproduce. But the *individual* insects of an ant colony don't integrate information and form a unified representation of the world, or of the problems they must solve. They make only simple decisions based on what they sense in their immediate environment. They are ignorant of the opportunities and threats posed by what's around them, and of their colony's goals and issues, and they receive no instructions regarding how to react. Instead, the ants' representation of the environment and its challenges are encoded in the colony. Countless interactions among individuals obeying simple preprogrammed rules result in the choices and behaviors of the colony as a whole, allowing it to thrive.

This is a classic example of bottom-up processing, in contrast to the "top-down" processing performed by organizations and programmed computers. As I've mentioned, our brains employ both. In top-down processing, the brain's executive structures orchestrate our reasoning, while bottom-up processing produces unorchestrated, elastic thinking.

Your Brain's Hierarchy

Our neurons are the "ants" of the human brain, producing the emergent phenomenon we call human intelligence. But we have 86 billion neurons, which is almost two hundred thousand times the number of ants in a typical colony. Also, unlike ants, which communicate with one or two other ants at any given time, each of our neurons is connected to thousands of other neurons via structures called axons and dendrites.

Due to that great complexity, the neurons in our brains possess several levels of organization. The brain appears superficially as a uniform mass of bulges and folds, but it is actually divided, and subdivided, into specialized regions. Neighboring neurons are connected into structures that perform specific functions, and those structures may themselves form part of larger structures, and so on—something like Russian nesting dolls.

At its largest scale, the outer layer of the brain's neural tissue is called the cortex. It is divided by a fissure into right and left hemispheres, and each hemisphere is divided into four lobes. Within each hemisphere, the frontmost lobe—the frontal lobe—is where the brain integrates information to produce thought and action. Like the other lobes, the frontal lobe is further subdivided. In particular, it contains the prefrontal cortex, one of the stars of this book.

Found only in mammals, the prefrontal cortex is the key structure that allows us to rise above the automatic response to environmental triggers that comes from scripted behavior.* Acting as the brain's "executive," it supervises our thought and decision-making by identifying objectives, guiding attention and planning, organizing behavior, monitoring consequences, and managing the tasks performed by other areas of the brain—a role analogous to that which a company's CEO performs.

The hierarchy goes on for several more levels. The prefrontal cortex, for example, is made of smaller structures, such as the lateral prefrontal cortex, an evolutionary advance found only in primates, which I'll talk about in chapter 9. The lateral prefrontal cortex, in turn, is made of yet smaller structures, such as the dorsolateral prefrontal cortex. And that structure, as I mentioned in the introduction, is itself made of a dozen substructures.

The structures at each level are interconnected in a complex manner, receiving input from some and providing its output to others. They are also connected to other structures that sit below the cortex, such as the substantia nigra, the ventral tegmental area, and the nucleus accumbens of the reward system. Each structure performs

---

* Birds' brains have an analogous structure.

tasks that contribute to the higher-level processing performed by the larger structures of which it is part. Ant colonies don't have that complex and hierarchical organization, and they don't complement their bottom-up processing with a degree of top-down control.

In humans, the executive brain helps us rise above the realm of purely habitual or automatic behavior by suppressing some thoughts and activating others. If your boss is unfairly berating you, it is your executive brain that allows you to save your anger for later, when you'll be sticking pins into the voodoo doll of your boss. Yet, in its attempt to suppress seemingly ill-advised or irrelevant ideas, your executive brain can impede original thinking. When we are at our best, our executive eases up enough that the brain achieves a balance of bottom-up and top-down operation. It is the balance of these modes of operation that determines the focus and breadth of your thinking.

That's the beauty of the human mind. We can execute an interplay of top-down and bottom-up processing, and of analytical thought and elastic thought. From that mix, ideas emerge that are organized and focused toward some end, many of them ideas that are not deducible using purely logical steps. We can program ourselves, we can create new concepts, and, best of all, we can alter our approach until we solve whatever problem the changing conditions of our environment have put before us.

An Intellectual Adventure

Our brains can operate in a top-down or bottom-up manner, but so can the individuals in organizations. Of all intellectual endeavors, academic science is one of the most bottom-up in the way it functions. Young scientists are invited to join research groups, but they're given a lot of freedom to follow their own ideas, rather than being dictated to from the top down by the group leader. That's especially true in theoretical physics, where the "startup cost" of pursuing a new idea isn't much more than the price of a pad of yellow paper. In the corporate world, bottom-up functioning is rare, and rigid, goal-directed thought is often valued over elastic thinking. Could corporations become "smarter" if they allowed for a greater degree of bottom-up processing?

One executive who believes the answer is yes is Nathan Myhrvold, who built the company Intellectual Ventures (IV). Myhrvold was a recent physics Ph.D. who for about a year had been working under Stephen Hawking when he took a summer leave of absence to start a business with some old school buddies. That leave turned into two years, and then the company was bought by Microsoft.

Myhrvold did well in Seattle, starting Microsoft's research division and staying until 1999. Just how well that worked out for him is clear from an exchange he and I had at his current lab, near Seattle. He proudly showed me an expensive precision miniature screwdriver set he had just splurged on. "I went back and forth on this quite a bit, because $250 is a lot to spend on tools like this," he said. "But I decided I could treat myself. After all, I own my own jet."

As Myhrvold told me this, he erupted with an enthusiastic and thunderous laugh. In his late fifties, jolly and cherubic, with a ruddy complexion, a sandy beard, and disheveled hair, he makes me think of a Santa Claus who's letting go after having had a few drinks. But this Santa's elves don't make children's toys. The scientists at IV, which he financed by leveraging connections he'd made at Microsoft, work on nuclear physics, optics, and food science.

Intellectual Ventures' goal is to create strange ideas that others don't think of, or that are considered misguided, and then license them. Myhrvold structured and developed the company to function like the human brain: a lot of interconnected people working together, with only minimal direction from above. That is why it is so interesting—it is perhaps unique as a corporation with bottom-up governance.

How did the bottom-up approach work out? Just look at IV's innovative products. A recurring theme at the company is finding novel uses for waste. One project seeks to turn the discarded outer hulls of coffee beans into an edible, gluten-free flour that could be mixed with ordinary flour to help feed the world's poor, an effort funded in part by Myhrvold's friend Bill Gates. The coffee flour would be a boon to impoverished countries for two reasons. First, the flour would cost just half as much as wheat flour, which for the most part has to be imported. Second, it would give coffee growers in the developing world a big bump in profits.

Consider the coffee you buy for $15 a pound. That translates to about $5 per pound to the coffee growers. But it costs, on average, $4.90 to grow that coffee, so they are making a profit of only ten cents. Myhrvold's company would haul the coffee bean hulls away, saving the growers five cents, and pay them an additional five cents, thus doubling the grower's profit while obtaining the raw material for the coffee flour at a cost that's low enough to make the final product much cheaper than flour made from wheat. Reworking coffee waste doesn't sound very sexy, but its impact could be enormous: Coffee growers generate billions of pounds of discarded hulls each year.

Another IV effort that raises eyebrows is called the Photonic Fence, an invention built around a laser that can shoot and kill flying bugs, à la Ronald Reagan's "Star Wars" missile defense system. It's aimed at reducing the incidence of malaria in Africa, as well as stopping the carnage winged insects impose on crops. The Photonic Fence is a quintessential example of the power of elastic thinking, the integration of wildly diverse ideas. First, from the mosquito experts, IV engineers learned that, late in the day, the insects fly toward the sunset but will stop and hover over a shadow or dark spot on the ground. Such areas are like a pickup joint for mosquitoes—where males encounter females and mate. Then there were the optics experts, who taught them about a technology called retroreflective coatings, which will shine light back directly at a source, no matter what angle it comes from. By setting up a retroreflective screen behind the mating area and aiming a low-powered targeting laser at it, the IV researchers are thus able to discriminate the shape, size, and wingbeat frequency of any bugs in the beam's path. That allows them to identify the species, and even the sex, of the bug—which is important in fighting malaria, because only the females are disease carriers. Finally, from laser experts they learned how to aim a higher-power laser at a targeted bug. In that manner, the apparatus can kill up to ten mosquitoes per second, using just the energy of a sixty-watt lightbulb.

IV does not manufacture any of its inventions. It makes money by buying, selling, and licensing patents like those related to the Photonic Fence. That can be controversial, because some say that reserving ideas at such an early stage stifles innovation. But the IV strategy has been working. Photonic Fence is now in the commercialization phase, coffee flour is already bringing in revenue, and IV has been spinning out, on average, a company every year. This is important, because it demonstrates the potential of applying what we have learned about the information processing in our minds to the way people are organized to attack real-world problems together.

Part III

Where New Ideas Come From

5

The Power of Your Point of View

A Paradigm Shift in Popcorn

David Wallerstein was not someone you would have thought of as a master of innovation. A young executive at the staid Balaban & Katz theater chain in the 1960s, he spent his days worrying about the bottom line in what was, even then, a marginal business. Then, as now, it wasn't the sale of film tickets that generated the bulk of revenue for a theater; it was selling salty popcorn and the sweet Coke to wash it down. Wallerstein, like everyone else, focused on increasing the sales of those high-margin concessions, and, like everyone else, he tried all the conventional profit-building tricks: two-for-one deals, matinee specials, etc. Profit remained flat.

Wallerstein was frustrated. He didn't understand what it would take to entice his customers to buy more. Then one evening he had an epiphany. Maybe people wanted more popcorn but didn't want to be seen eating two bags of it. Perhaps they feared that buying a second bag might make them appear piggish.

Wallerstein decided that if he could find a way to circumvent their aversion to buying a second bag, it would help his profitability. It was easy. Just offer a larger bag. And so Wallerstein introduced a new size of popcorn to the moviegoing world: jumbo. The results astonished him. Not only did popcorn sales immediately shoot upward, but so did sales of that other high-profit treat, Coca-Cola.

Wallerstein had unearthed what is a basic law in the food industry today: People will gorge themselves on enormous quantities of food if "enormous" is one of the serving sizes offered. In the Bible, gluttony is a sin, but apparently people consider a restaurant menu to be a higher authority, and if it offers an eight-scoop banana split, that gives permission.

Economists write many scholarly articles, usually starting with the assumption that people act rationally, which in reality excludes everyone except those with certain rare brain disorders. Wallerstein, on the other hand, uncovered a truth about *actual* human behavior. Did the food industry give him a trophy for the new idea and adopt the Wallerstein strategy? No.

In his classic book *The Structure of Scientific Revolutions*, Thomas Kuhn wrote about what he called "paradigm shifts" in science. These are alterations in scientific thinking that represent more than incremental advances. They are alterations of the framework of thinking, of the set of shared concepts and assumptions within which scientists do their theorizing (until the next paradigm shift). To solve problems and draw conclusions within an existing framework requires a blend of analytical and elastic thinking. But the act of envisioning a new framework for thought relies heavily on the elastic component—on skills such as imagination and integrative thinking.

Paradigm shifts are peculiar in that they leave many previously successful people behind, people whose rigidity of thought causes them to cling to the old framework to which they are accustomed, despite often overwhelming evidence that the paradigm shift is valid. Or sometimes, those who cannot accept a shift form the vast majority, and its implementation is blocked or delayed. That was the fate of Wallerstein's ideas.

Wallerstein's approach to selling snacks represented a paradigm shift for the food industry—though it seems obvious today, it was heresy back then. In the 1960s, people viewed consuming large amounts of food as unattractive, and executives couldn't accept the idea that with their nudging, that might change—that it was simply the act of having to purchase the second helping that stood in the way of unbridled consumption. What's more, many food executives saw larger portion size as a form of "discounting," an act that conventional wisdom told them would hurt the view of theirs as a quality brand. As a result, Wallerstein's innovation didn't catch on elsewhere.

Even when Wallerstein himself landed at McDonald's in the mid-1970s, nothing changed—he couldn't convince Ray Kroc, the McDonald's founder, to introduce a larger size of French fries. "If people want more fries," Kroc said, "they can buy two bags." McDonald's continued to resist but finally adopted the strategy in 1990. By then, what had come to be called "supersizing" had become the new conventional wisdom. But it had taken the food industry longer to recognize the law of human gluttony than it took the physics community to embrace the theory of relativity. In hindsight, the mental adjustment to a framework of thought in which large portions are the standard seems easy. So does the idea of chocolate chip cookies, now that someone has invented them.

The Structure of Personal Revolutions

In *The Structure of Scientific Revolutions*, Kuhn wrote that scientists hold institutionalized everyday beliefs, which may, on occasion, be altered by a transformational discovery. But that is also true of us in our personal lives. We each develop our point of view on common issues during our first few decades of life or our first years in a new job. We form a framework to apply those ideas and apply them when we're called upon to draw conclusions in those realms. For some, those paradigms never evolve, but for the fortunate they do change, often in Kuhnian jumps. Those who are open to such personal paradigm shifts—to altering their attitudes and beliefs—have always had an advantage in life, because they are more able to adapt to changing circumstances. In today's society, that is especially important.

To aid my ability in that regard, I sometimes engage in a little mental flexibility exercise. I list some of my strongly held beliefs on slips of paper. I fold them, pick one, and imagine someone telling me that the belief written on it is false. Of course, I don't for a moment think my belief is really wrong. That is just the point—when my instinct to reject the notion that I'm wrong kicks in, that's when I am in the position of all those who, in the past, failed to adapt to challenges to the ideas that they, too, held firmly.

That's when I push myself, and try harder to be open to the possibility that I'm mistaken. Why do I hold that belief? Are there those who don't? Do I respect them, or at least some of them? Why might they have come to a different conclusion? I try to recall times in the past when I *was* wrong about something, even though I'd been confident of being right. The bigger the mistake, the better. The process helps me to understand that the mental adjustment to a new thought paradigm isn't as easy as hindsight always makes it seem.

The exercise led me to challenge myself, for example, on the question of immigration. My parents emigrated from Poland after "the war," as it was called in my household. All their friends were also immigrants, and Holocaust survivors. When I started school, I could distinguish a Hungarian from a Czech, but I had never met an adult who was a natural-born American. I thought it was normal to eat brisket on Thanksgiving, and I was sent to speech therapy because I spoke with a Polish accent that my teachers mistook for a speech impediment.

Owing to that background, I've always been in favor of our country's accepting the tired and poor, and, after that, if there is room, the powerful and rich. I want to grant to others the opportunities that my family was given. I've felt anger toward those who feel otherwise, especially when, during the 2016 presidential election campaign, the talk turned to building a wall along the Mexican border.

That was how matters stood when, in one of my mental flexibility exercises, I picked a slip that read: *The supporters of building a wall with Mexico are evil.* I remember rolling my eyes—there was no way I was wrong about this one. But I dutifully put on my scientist's cap and tried to examine the argument for the wall as if it were a scientific issue, devoid of human relevance. At first I pondered all the disputes over the data on the contributions of immigrants, or the efficacy or cost of a wall. But then I decided that those were side issues. My belief was rooted not in such data, but in my feeling that a wall was simply an affront to what I wanted this country to stand for.

What would all those evil people on the "other side" say to that? I started to watch Fox News to find out. Almost hidden in all the noise was what I concluded to be their basic logic: We have laws about immigration. If we don't like them we should change them, but as long as we have them, if they are ineffective, it makes sense to consider new means of enforcing them. I realized that if you find that logic compelling, it does not necessarily mean that you are the type who kicks dogs and pulls the wings off flies.

We tend to make quick initial assessments of issues based on the assumptions of the paradigms we follow. When people challenge our assessment, we tend to push back. Whatever our politics, the more we argue with others, the further we can dig in, and sometimes vilify those who disagree. Then we reinforce our ideas by preaching to the choir—our friends. But the mental flexibility to consider theories that contradict our beliefs and don't fit our existing paradigms not only can make you a genius in science; it is also beneficial in everyday life.

In the business world, accepting challenges to old ways is equally important, because industries are quickly evolving. Apple, for instance, is a corporation that makes and sells products. As a result, it is classified by the U.S. government as a manufacturing company. But that classification is based on what has become an outdated way of thinking. For, though the company gets the bulk of its revenue from selling physical products, virtually all of Apple's products are manufactured for it by others. Adopting a twenty-first-century model, Apple thus avoids having to invest in factories and is better positioned to change directions nimbly, gaining an advantage over less forward-thinking competition.

Or look at Nike. It is heading toward what, until recently, seemed a science-fiction-like manufacturing method: 3-D printing. Within the company, they call the initiative the "manufacturing revolution." Partnering with HP in 2016, Nike has already employed the technology for prototyping new designs. And it is looking forward to a future in which 3-D printing and 3-D knitting are combined to create shoes on the spot, in-store, and custom-tailored to the precise measurements of each person's feet. Like Apple, Nike presents an existential challenge to any competitor that follows the old principle that one should not question assumptions and methodologies that have, in the past, led to success.

Reimagining Our Frames of Thought

I was always surprised, when attending a Christian church service, at how quiet the worshippers were. We Jews like to talk. And so at synagogues, the rabbi is often in the position of having to pound his or her fist on the podium to quiet the din. One time, in a sermon, the rabbi addressed this. "If you would ask me if it is acceptable to socialize with your neighbors while praying, I would say I'd rather you didn't. You are here to worship, and your banter is distracting, if not disrespectful," he said. "But if you were to ask me, is it okay to go to the synagogue and pray while you socialize with your friends, I'd say, 'Certainly! We are always happy to have you.'" He went on to a lengthy discussion of Talmudic principles, the kind of microscopic dissection of issues you get used to if you attend synagogue. But to me, the point was that the way you frame an issue has a profound influence on the results of your analysis.

Consider these brainteasers from a 2015 study in *The Journal of Problem Solving*. To solve them, like Wallerstein, you will have to question your assumptions and alter your framework of thinking. If you enjoy riddles, give them a try:

1. A man is reading a book when the lights go off, but even though the room he is in is pitch-dark, the man goes on reading. How? (The book was not in an electronic format.)
2. A magician claimed to be able to throw a Ping-Pong ball so that it would go a short distance, come to a dead stop, and then reverse itself. He added that he would not accomplish this

by bouncing the ball off any object, tying anything to it, or giving it spin. How could he perform this feat?

3. Two mothers and two daughters were fishing. They managed to catch one big fish, one small fish, and one fat fish. Since only three fish were caught, how is it possible that each woman caught her own fish?

4. Marsha and Marjorie were born on the same day of the same month of the same year to the same mother and the same father—yet they are not twins. How is that possible?

In the study I quoted, each of these brainteasers was solved, on average, by fewer than half of the subjects. How did you do?

The reason the riddles are difficult is that each of them suggests, in most people's minds, a certain picture:

1. A man staring into a book.
2. A man tossing a Ping-Pong ball onto a table or the ground.
3. A group of four women.
4. A pair of twins, Marsha and Marjorie.

These pictures determine our framework of thought as we try to find the answers. As long as we adhere to them, the ideas that our associative brains pass to our consciousness will be consistent with them. But these pictures are incorrect interpretations of the circumstances described by the riddles. To solve the riddles, these preconceptions must be abandoned.

What makes riddles so difficult is that they are deliberately designed so that the wrong interpretation will come to mind automatically, with little or no conscious consideration. It's the interpretation our brains deemed the most likely to be appropriate based on past experience, a hidden assumption we are unaware of making—but it is a picture that is inconsistent with the novel situation envisioned in the riddle. Like many challenging problems, the riddles are made difficult not by what we don't know, but by what we *do* know—or think we know—which turns out to be incorrect.

Take the first riddle. In the great majority of circumstances that most of us encounter, a man who is reading a book is indeed staring at its pages. But though that is one possible scenario described by the riddle, as we'll soon see, there is another possibility, and to realize this and let go of that initial picture is the key to success. That's analogous to the dynamics of paradigm shifts in business and science. In those fields, changing circumstances invalidate assumptions that are so ingrained that people don't question them, or have trouble accepting that they no longer hold. Success comes to those who realize this and can revise their understanding of the situation.

Here are the solutions to the riddles. In the first, the man did not require light to read because he was blind, and was reading the book in braille. In the second, the magician threw the ball upward into the air, not horizontally, so its motion was reversed by gravity, not by a collision with the ground, a table, or a wall. In the third, only three fish were caught because the two mothers and two daughters constitute only three women—a girl, her mother, and the mother's mother. And in the fourth, Marsha and Marjorie did not represent the entire brood—they weren't twins, but triplets.

We encounter many challenges in life. We know how to deal with some because we have bumped into them before. Others are novel but can be overcome through straightforward analytical thinking. But some problems have withstood our attempts at solution. Often, as in these riddles, that is because no solution exists within the framework in which people have thought about them—but a solution could be found if one adopted a new point of view.

When we talk about triumphs of intellect, we tend to focus on brilliant analytical thinking, the kind of thought produced by powerful logic. But rarely do we recognize the contribution of being able to reimagine the framework in which our thinking occurs, the terms in which our mind defines the issue we are considering. That's the product of elastic thinking, a task that requires that squishy ability called "judgment." Creating new representations is difficult to automate, and most animals have trouble doing it, but it is often the key to successful problem-solving in the human world.

The Dog-and-Bone Problem

In this age, issues that require us to alter our framework of thought are more common than ever before. That's what disruptive change is all about—it is change that demands new paradigms and different ways of thinking. Psychologists call the process of altering the framework through which you analyze an issue "restructuring." That most fundamental operation of our minds often spells the difference between finding an answer and reaching an impasse. Or, once you've reached an impasse, restructuring is often the only way to overcome it. Today, as the assumptions of the past are being rendered obsolete at a blinding pace, the ability to restructure your thinking is less and less a requirement for standout achievement and more and more simply a requirement for survival.

Computer scientist Douglas Hofstadter illustrates the importance of restructuring with what he calls the "dog-and-bone problem." Imagine you are a dog, and a kind human has just tossed you a bone, but it landed in the neighbor's yard, just on the other side of a ten-foot-tall chain-link fence. Behind you is an open gate; before you is a tasty snack. Your mouth waters as you look at the bone, but how do you get to it?

Unless they have encountered this problem before, most dogs will represent the situation in a strict geographical sense. They make an internal map of their own location and the bone's; they have an idea of the distances on that map; and they have a goal of moving, over time, to diminish that distance. The dog might start out thirty feet from the bone. As the dog moves toward it, that distance will decrease, and the dog concludes from its innate programming that when the distance gets to zero it will have reached its target.

A dog—or a robot—with that program will run toward the bone until it encounters the fence, at which point it will have reached an impasse. Its distance from the bone may have diminished to just a matter of inches, but it can proceed no farther. Some dogs will then just stare at the bone and bark in frustration, or roll over and wait for you to pet their belly. Others, possessing the concept of digging as a way of traveling under obstacles, might try doing just that. But some bright dogs will possess enough mental elasticity to change the framework through which they think of the situation: They will realize that their physical distance from the bone is not the same as their distance from their goal.

Standing at the fence, these dogs realize that, although they are just a few inches from the bone, they are far from being able to reach it. And so they will change the notion of distance they employ for the purpose of this problem. They will understand that, although they are standing physically near the bone, the open gate is closer to the bone, in the sense of achieving their goal, than they are. So rather than employ literal geometric distance as their gauge of progress, they will use a definition of distance in what cognitive scientists call "problem space."

In this case, the distance in problem space is the distance *along a path that will take them to the bone*. In problem space, if the dog begins by moving toward the bone, it will be increasing its distance from its goal, but if it moves toward the open gate, it will be diminishing it. And so dogs that create this new framework will race to the open gate.

Solving the dog-and-bone problem, once it is effectively framed, is easy. But realizing that a new framework is needed, and then creating one, requires elastic thinking. Effective thinking often boils down to that—the ability to restructure your framework of thought about the facts and issues. And so the dog-and-bone problem, though simple, separates the thinkers from the nonthinkers, the humans and bright dogs from the chess-playing computers.

How Mathematicians Think

If there is one field whose bread and butter is restructuring, and that can therefore teach us much about innovation and creative thought, it is mathematics. Most of us have no idea how mathematicians think, but we can learn a lot from their deftness at creating alternative frameworks for difficult problems.

Take this problem, which is really a mathematics problem but masquerades as an everyday riddle: You have an eight-by-eight checkerboard and thirty-two dominoes. Each domino can cover two horizontally or vertically adjacent squares, and it is easy to see how one might place them in a manner that covers all sixty-four. Now imagine that you toss away one domino and remove two squares, from two diagonally opposite corners. Can you cover the remaining sixty-two squares with thirty-one dominoes? Whether your answer is yes or no, explain how you know that. No domino is allowed to "stick out" past the boundary of the board.

When presented with this problem, most people try various arrangements of the dominoes and then, after failing, suspect that covering the entire board is impossible. But how to prove it? Trying one failed configuration after another won't do it, because there are too many possibilities.

The "mutilated checkerboard" riddle is a kind of human upgrade of the simple dog-and bone problem. There is an easy answer, but it involves looking at the issue within a new framework, restructuring the question in a manner that abandons the literal attempts to cover the board and instead formulates the problem in a new way. How?

The key is this: Instead of framing the problem as a search through the "space" of ways to cover the checkerboard with dominoes, frame it instead in terms of a search through the space of "laws" that govern the act of placing dominoes on the board. Of course, first you'll have to formulate those laws. Here is one: *Each domino covers two squares.* Can you think of others? Once you've identified all the laws you can think of—there aren't many—look at the issue of whether you can cover the entire mutilated board in the context of those laws. You'll find that there is a law that would have to be violated in order to cover the entire mutilated board, and so the answer is no, you cannot.

If you thought of the following law, you probably solved the riddle: *Because each domino covers two adjacent squares, each domino, when placed on the board, covers one black square and one white square.* That law means that there is no way to place dominoes on the board in a manner that covers an unequal number of black and white squares. The complete checkerboard has an equal number of white and black squares, so this law doesn't preclude you from covering it with dominoes. But the mutilated board, with two opposite corners removed, has *thirty-two* white squares and *thirty* black squares (or vice versa), so the law tells us that there is no way to cover it.

The annals of mathematics, and much of problem-solving in all fields, can be viewed as a relentless series of assaults on unhelpful frameworks, employing the weapon of restructuring. Here's an example from real mathematics: What is the solution of the equation $x^2 = -1$? Since the square of any number is a positive number, asking someone to solve that problem seems to be like saying, "You have two pounds of flounder and a carrot. How do you make beef stew?" For centuries, mathematicians assumed there was no answer. But they were all working within the framework of ordinary mathematics, which we today call the "real numbers."

In the sixteenth century, the Italian mathematician Rafael Bombelli realized that the fact that the square root of -1 is not a number we can, say, count on our fingers doesn't mean it is a number that cannot be useful to our minds. After all, we use negative numbers, and those don't correspond to a number of fingers or to any physical quantity. Five hundred years ago, that was Bombelli's great restructuring: to look at numbers as abstractions that obey rules, rather than as concrete entities. And so Bombelli questioned whether there might be some legitimate mathematical framework of numbers that allows for the square root of −1, without regard for whether such numbers could be used to count or measure things.

Bombelli probed the issue by saying: Suppose such a number *does* exist. Does that lead to a logical contradiction? And if not, what would that number's properties be? He found that a number satisfying $x^2 = -1$ does *not* lead to a logical contradiction, and he successfully discovered some of its novel properties. Today we write Bombelli's number as i, and call it an *imaginary* number.

Imaginary numbers are a fundamental cornerstone of many fields of mathematics, and they play a critical role in most of physics. For example, they are the natural way to describe wave phenomena; thus, without imaginary numbers, we would probably have no quantum theory and, hence, no electronics—and so the modern world as we know it would not exist.

Imaginary numbers are now taught as elementary math. Advanced high school students have no trouble learning what the most advanced medieval scholars could not fathom and many could not accept, because, like the idea of jumbo sizes, it contradicted the paradigm of conventional thought.

The Influence of Culture

The stories of Wallerstein and Bombelli couldn't be more different, yet they both illustrate that one important influence on our ability to arrive at new representations comes from outside ourselves—from our professional, social, and cultural norms. They can be the norms of our family, our peers, our country, our ethnicity, our field of expertise, or even the particular company we work for. We tend to think that national and ethnic culture exerts the biggest influence on an individual's thinking, but if you know any mathematicians, they probably think quite differently from the lawyers you know, who think very differently from the chefs, accountants, police detectives, and poets you know, and those differences can be just as great.

Whatever its source, the influence of culture is so powerful, it affects even our perception of physical objects. Consider a recent study by University of Michigan psychologist Shinobu Kitayama and colleagues, who studied the differences in how European American and Japanese American subjects perceive simple geometric figures.

Culture is to a group what personality is to an individual. European culture, psychologists have found, stresses independence and literal thinking, while the Japanese culture is more communal and stresses situation and context. To investigate the cognitive consequences of those differences, in one set of experiments Kitayama showed his subjects a "standard" box drawn on a sheet of paper, with a line segment precisely one-third the height of the box drawn vertically down from its top border, as in the illustration below.

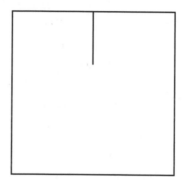

The subjects were also given their own sheet of paper, which had an outline of a similar box on it. That box was a *different size* than the standard box and had *no line* hanging down from the box's top.

Each subject was given a pencil and asked to replicate the vertical line of the standard box within his or her box. Some were asked to draw a line of the same *length* as on the standard; others were asked to draw a line of the same *proportion* (one-third of the height) to the box that framed it. Those two requests differed in an essential way. In the former, one can ignore the box, while in the latter the relation of the box and the line are paramount.

The researchers designed the study around this difference because the box is the context for the line, and context is the element that is stressed in Japanese culture. Kitayama thus predicted that the Japanese would do better than the Europeans when asked to match proportion, but not when told to match the line by length—and that is exactly how the experiment turned out.

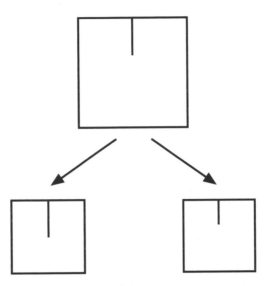

Kitayama's boxes. Left: line length matches the original. Right: line proportion matches the original.

Kitayama's study probed how people think in an artificial laboratory setting, but cultural differences that reach as deep as one's physical perception are bound to profoundly affect the way different people approach problems in their respective societies. So sociologists asked: Does culture affect a society's level of innovation? If so, the relative ranking of countries with regard to innovation should be stable over time, reflecting each society's underlying culture.

The table below displays the results of one study that addressed that issue. It ranks the United States and thirteen European countries of roughly comparable wealth with regard to number of inventions patented per capita over the decade 1971–1980. It shows that most retained their same rank over the entire decade.

	RANK IN 1971	RANK IN 1980	10-YEAR CHANGE
Switzerland	1	1	0
Sweden	2	2	0
USA	3	3	0
France	4	7	−3
Great Britain	5	6	−1
Austria	6	4	2
Belgium	7	10	−3
Germany	8	5	3
Norway	9	9	0
Finland	10	8	2
Spain	11	11	0
Denmark	12	12	0
Netherlands	13	13	0
Portugal	14	14	0

The study was no fluke. For example, the table below shows analogous rankings obtained by other researchers, and for a different decade, 1995–2005. In this study, only a subclass of inventions was studied, so the two tables are not directly comparable, but what is important is that the rankings are similar, and they are also stable over time.

	RANK IN 1995	RANK IN 2005	10-YEAR CHANGE
Switzerland	1	1	0
Sweden	2	2	0
Finland	3	6	−3
Germany	4	3	1
Netherlands	5	4	1
USA	6	5	1
Belgium	7	10	−3
Denmark	8	7	1
France	9	9	0
Austria	10	8	2
Great Britain	11	11	0
Norway	12	12	0

Ireland	13	13	0
Spain	14	14	0
Portugal	15	15	0

Our culture can provide an approach to issues that will help us see solutions, but it can also hinder us. A strong cultural identity, if it produces a deeply ingrained approach to problems, can make it more difficult to change that approach, even if it is not working. On the other hand, exposing yourself to other cultures is beneficial, because those who grew up in or work within a different culture will often have different attitudes toward life, and studies show that simply interacting with such people can open our minds and increase our elasticity of thought. The broader perspective that such exposure brings makes us more likely to break our old patterns and free ourselves from rigid thought patterns that may be holding us back.

But whether we need a new framework to shape our thinking or can find the solution within our existing structure of thought, where do the ideas we seek come from? As the next chapter explains, the process of idea generation is seated deep within our unconscious mind and is most active when our conscious processes of analytical thought are at rest.

6

Thinking When You're Not Thinking

Nature's Plan B

Lying in her bed in the village of Cologny, near Lake Geneva in Switzerland, Mary Godwin felt frustrated. It was past two in the morning, another dreary night in a dreary, rainy month of June. The incessant rain was nothing new to her: She'd grown up in London and had also spent considerable time in Scotland. But on this night, the gloom of the weather reflected her mood.

Mary, a pale, dainty woman with auburn hair and deep hazel eyes, was being hard on herself. The year was 1816, and she was only eighteen years old. She had come to spend that summer in Switzerland with her half sister, her friends, and her lover. Late one night during a particularly wild downpour, they had all gathered around a log fire to read aloud from a volume of ghost stories, and then decided that they'd each write one.

By the next night, everyone but Mary had their stories. Days passed. The others kept asking, "Have you thought of a story?" and she kept having to reply with "a mortifying negative." She began to feel unworthy of the intellectual company of her friends and lover. The insecurities only fed her frustration.

Mary's friends continued their late-night fireside ruminations, and on the night in question they spoke of the "nature and principle of life." They mused about some experiments by Erasmus Darwin, in which he had supposedly "preserved a piece of vermicelli in a glass case, till by some extraordinary means it began to move with voluntary motion." On reading those words, my reaction was that we've all had leftovers like that. But this was an intellectual group, and they wondered, *Could life be created so simply, and by what force?* Sometime around midnight, they all went to sleep—all but Mary, that is, who lay in bed, staring at the ceiling. She couldn't sleep, but she shut her eyes and resolved to maintain a quiet mind. It was time to take a pause from her efforts at creating a story.

It was while Mary was in that "relaxed" state of mind that the outline of the story she had been seeking suddenly came to her. Apparently inspired by the night's musings, she would recall, "my imagination, unbidden, possessed and guided me." Mary said, "I saw—with shut eyes, but acute mental vision . . . the pale student of unhallowed arts kneeling beside the thing he had put together." Mary Godwin—who would become, after marrying her lover, Mary Shelley—had had the vision that in 1818 would lead to her book *Frankenstein; or, The Modern Prometheus.*

Every creation begins life as a challenge, just as every answer begins as a question. As we saw in chapter 3, there is a lot in common among the desire to paint a canvas, solve a problem, invent a device, produce a business plan, or prove a point in physics. What these endeavors also share is that, if we endure the same intense discomfort and frustration that Mary Shelley felt, from deep within the recesses of our elastic mind, an idea may suddenly come.

The elastic thinking that produces ideas doesn't consist of a linear train of steps, as analytical thought does. Sometimes big, sometimes inconsequential, sometimes in crowds, sometimes as loners, our ideas seem to just appear. But ideas don't come from nowhere; they are produced in our unconscious minds.

To Mary, the mode of thought that led to her first rough but inspired vision of *Frankenstein* was shrouded in magic and mystery. How could the story, which she had struggled for days to create, come to her as she rested in bed, thinking about nothing in particular?

Before the advent of neuroscience and the technology that made neuroscience possible, it was enormously difficult to understand how a daydream or a wandering mind could produce answers when our conscious efforts to do so had failed. But today we know that quiet brains are not idle brains, that in periods of mental peace, our unconscious may be overflowing with activity. Today, two hundred years after the origin of *Frankenstein*, we can measure and monitor the physical underpinnings of that activity. We understand that, magical as it may seem to be, thinking while we are not consciously focused is a fundamental feature of the mammalian brain, possessed even by lowly and primitive rodents. Known as the brain's *default mode* of thought, it is a key mental process in elastic thinking.

The Dark Energy of the Brain

Marcus Raichle calls what he has been studying for the past twenty years "dark energy." In astrophysics, the term *dark energy* refers to something mysterious that permeates all of space and constitutes two-thirds of all the energy in the universe yet goes unseen in everyday life. As a result, it went unnoticed through centuries of astronomy and physics, until it was discovered by accident in the late 1990s. But Raichle is a neuroscientist, not an astronomer, and the energy he has been studying is the "dark energy" of the brain—the energy of the brain's default mode.

The analogy is apt because, like the astrophysicist's dark energy, the dark energy of the default mode is a kind of "background" energy—it is energy that arises from the background of cerebral activity. And it, too, despite being substantial, was long hidden from us—because the default mode is not activated during everyday activity. Instead, it becomes active when the executive brain is not directing our analytical thought processes to anything in particular.

There is currently an explosion of research on the default mode, the result of a series of papers Raichle wrote when just a few years into his work on the topic, in 2001. As I write this, his initial article has been cited more than seven thousand times—an average of more than one new scientific paper on the subject each day, each of them the product of months or years of work. But, like many scientific breakthroughs, the concept of the default mode swam unheralded in that great sea of scientific ideas long before Raichle rediscovered it and published the paper that brought it to the level of prominence it has today.

The tale begins in 1897, when a twenty-three-year-old fresh out of graduate school took a position in the psychiatry clinic at the University of Jena, in Germany. His specialty was neuropsychiatry. With roots in the seventeenth-century work of Thomas Willis, it is the study of how one might connect mental disorders to specific processes in the brain. The only way to observe those processes in 1897 was to saw open a skull, so people weren't exactly flocking to that field. But that young psychiatrist would change that, while working in Jena for the next forty-one years, and creating the first great technological tool of neuroscience.

Hans Berger's colleagues described him as shy, reticent, inhibited, brooding, detail-oriented, and highly self-critical. One said he was "obviously fond of his instruments and physical apparatus and somewhat afraid of his patients." Another, who would later serve as a subject in Berger's experiments, said that Berger would never "take any step that was not in accordance with [his] routine. His days resembled one another like two drops of water. Year after year he delivered the same lectures. He was the personification of static."

And yet Berger had a secret and audacious interior life. In his diary, he made wildly unorthodox scientific speculations. He interspersed them with original poetry and spiritual reflections. And in his research, which he kept secret from virtually everyone, he pursued what were, for his day, shocking scientific ideas. One was connected to an experience he had had when he was twenty years old and serving in the military.

During a training exercise, Berger had been thrown from his horse and narrowly escaped death. That same evening, he received a telegram from his father—the first he had ever gotten from anyone in his family—inquiring about his health. As he later learned, his sister, who lived far away, had urged their father to contact him, because that same morning she had been struck by a sudden attack of apprehension about his safety. The confluence of events convinced Berger that somehow his own terror had been communicated to his sister. As he wrote many years later, "It was a case of spontaneous telepathy in which at a time of mortal danger, and as I contemplated certain death, I transmitted my thoughts, while my sister, who was particularly close to me, acted as the receiver." After that, he became obsessed with trying to understand how the energy of human thought can be transmitted from one person to another.

Today the concept of mental telepathy sounds unscientific, because it has long since been thoroughly investigated and discredited; in Berger's era, however, the evidence against it was far thinner. In any case, what ultimately defines the value of an investigation in science is not what is being investigated, but how carefully and intelligently the research is done. Berger went about his research with the same strict scientific rigor that his colleagues always attributed to him. But in order to bring that rigor to an understanding of the energy transformations in the nervous system, and correlate them to mental experience, he had to find a way to measure the brain's energy.

Although no one had ever tackled that problem before, Berger had a brilliant idea about how to do it. Inspired by the work of Italian physiologist Angelo Mosso, Berger reasoned that since metabolism requires oxygen, he could measure blood flow as a proxy for energy. That principle was almost a hundred years ahead of its time—it is the key to the functional magnetic resonance imaging technology (fMRI) that helped launch the neuroscience revolution in the 1990s. Of course, fMRI depends on massive superconducting magnets, powerful computers, and a theoretical design based on quantum theory, none of which were available to Berger when he began his investigations early in the twentieth century. All Berger had to work with were items of the sort you might find today in a middle school physics lab, plus a saw. How could he observe blood flow in the brain with that?

The answer is gruesome, but that was where Berger got lucky: The Jena clinic where he worked put him in regular contact with patients who, due to either a tumor or, often, an equestrian accident, *had* to have a bit of their skull removed in the course of their treatment. One man's ceiling may be another man's floor, but here, one man's "craniotomy" was another man's window to the brain.

Berger's first experimental subject was a twenty-three-year-old factory worker with an eight-centimeter-diameter gap in his skull, the result of two surgical attempts to remove a lodged bullet. Though the man suffered from intermittent seizures, he was cognitively unaffected. With the man's permission, Berger fashioned a small rubber bladder, filled it with water, and affixed it snugly within the hole in the man's head. He connected the bladder to a device designed to record changes in its volume—when blood flowed to the area of the brain beneath the bladder, the brain would swell slightly and compress it.

Berger asked his patient to perform tasks such as simple mental arithmetic, counting the spots on the opposite wall, and anticipating being touched across the ear by a feather. He called the thoughts required for those tasks "voluntary concentration," and he measured blood flow to the brain as his patient performed them. Berger also measured blood flow due to "involuntary attention." His protocol for that was not as innocuous: He stepped behind his unsuspecting subject and fired a gun.

If the field of neuropsychiatry had a code of ethics back then, the bar must've been set pretty low. In addition to being hard on the patients, Berger's experiments were plagued by technical problems. Over the years, they led to some publications—such as a 1910 book, *Investigations on the Temperature of the Brain*—in which he argued that the chemical energy of the brain can be converted into heat, work, and electrical "psychic energy." But his conclusions—and his data—were weak, and he was battered with self-doubt and fighting depression.

By 1920, Berger had gotten bolder. He explored brain function by inserting an electrode into patients' brains in order to deliver an electrical current. The plan was to correlate brain geography with what the subject experienced when various cortical locations were stimulated by the weak current. He was conducting such experiments on the brain of a seventeen-year-old college student in June 1924 when he had an epiphany: Why not remove the electrodes from the cortical *stimulator* and attach the electrodes instead to a device used to *measure* electrical current? In other words, he turned the tables—rather than send current into the brain, the new setup would allow him to study the brain's own electricity.

That proved to be the key to Berger's success, because, over the next five years, he learned to make those readings from outside the skull, by attaching electrodes to a subject's scalp. As one might imagine, that greatly increased his pool of volunteers. It was a tool that could be used on anyone, and indeed, he made thousands of readings, including many on his own son.

Berger called his device the electroencephalograph, or EEG. In 1929, at the age of fifty-six, Berger finally published his first paper on that research, "On the Electroencephalogram of Man." In the next decade he would publish fourteen more papers, each with the same title, distinguishing them only by number.

Berger's EEG was one of the most influential inventions of the twentieth century. It opened a window to the brain, enabling neuropsychiatry to become a true science. Today, scientists regularly employ the EEG to study mental processes like those that transpired in Mary Shelley's brain that night as she relaxed her mind. But Berger himself made the first big discovery in that regard.

Employing his new device, Berger demonstrated that the brain is active even when a person is *not* engaging in conscious thought, when the mind is daydreaming or wandering—as Mary Shelley's was when her ideas came to her. Even more unexpected was the fact that the electrical energy characteristic of that inactive state, as measured by the EEG, subsided the moment voluntary concentration began or if a subject's attention was drawn to some event in the surrounding environment.

Berger's ideas were contrary to the scientific wisdom of the day, which held that the brain was electrically active only during tasks that demanded attention. He preached about the importance of his new discovery, but few listened. Scientists knew that when people were not thinking, there must be some residual brain activity to enable functions such as breathing and heartbeat to continue, and they assumed that whatever else Berger's EEG detected was just random noise. That view was not unreasonable, and yet, had others been more open-minded, they'd have realized, as Berger did, that the signal was not random. Sadly, this was an instance of an existing paradigm getting in the way of intellectual progress—an all-too-common story.

By the late 1930s, Berger's work on the EEG had spawned a huge field, but no one studied the energy of the resting brain. Contemporary research had progressed in other directions, and Berger was left behind. Then, on September 30, 1938, while making the rounds in his clinic, Berger was abruptly called to the telephone and told by the Nazi-approved authorities that he was being fired the next day. Shortly thereafter, his laboratory was dismantled.

In May 1941, with World War II in full swing, his career halted by the Nazis, and the field of EEG research not moving in the direction he wanted, Berger wrote in his diary, "I have sleepless nights in which I keep brooding and struggling with self-accusations. I am unable to read or work in any organized way, but I want to force myself, for like this it is unbearable."

His career effectively over, Berger felt he had not been successful at his lifelong goal of connecting the electrical processes of the brain to what is experienced in the mind. He'd made a major step along that path with his discovery of the electrical energy of the brain during daydreaming, but he'd been unable to carry it further or to convince anyone of its importance. The final words of Berger's last published paper were a plea for his colleagues to take that idea seriously:

> I would like to draw attention to a certain point I have made in the past. When mental work is performed or when the type of activity designated as active conscious activity becomes manifest in any way . . . a considerable decrease in the amplitude of the potential oscillations of the human brain occurs in association with this shift in cortical activity.

Berger could just as well have screamed into the vacuum of space. His words reached no one—one of the costs of being so far ahead of your time. On May 30, 1941, Hans Berger took his own life. On his study wall hung a poem written by his maternal grandfather, poet Friedrich Rückert:

> *Each man faces an image*
> *Of what he is meant to become.*
> *As long as he does not achieve it*
> *He cannot achieve his full measure of peace.*

The Symphonies in Idle Minds

As I speak with her, Nancy Andreasen, a brunette with close-cropped hair, is approaching eighty. She's a medical doctor with a Ph.D. in English. That's not the usual combination you find in neuroscience—or anywhere in science. The Ph.D. came first and led to a job as a professor of Renaissance literature at the University of Iowa. Then, one day while she was lying in bed for a week after a difficult pregnancy and delivery, Andreasen's mind wandered, and she had a life-altering idea—a sudden realization that she wanted to make a change.

As Andreasen recounted this to me, I thought of Mary Shelley, and how the story of Frankenstein had come to her. Except that in Andreasen's case, the story she was creating was a rewrite of her own life. At the time of her revelation, she had just had a book about the poet John Donne accepted for publication by Princeton University Press—which for most English professors in the early days of their career would have been a great triumph. Not for her, however. "I realized I wanted to do something that changed people's lives more than writing a book on John Donne," she said.

To find something that would change people's lives more than a book on John Donne is not a very tall order. A glass of chardonnay would do it for most of us. But the "something" Andreasen chose was very ambitious. She decided to go to medical school and study neuropsychiatry—Berger's field. This was quite a radical step for someone who, as an English major, had taken very few science or math courses in college. She would have to build her new career from scratch. She would also have to do it at a time when women faced far more barriers than they do today.

This was the late 1960s. As a high school student, Andreasen had had to turn down a prestigious scholarship to Harvard because her father didn't think a young girl should stray that far from home. As an academic who wanted to publish in scholarly journals, she had learned that, to be taken more seriously, it was best to hide her gender by using initials for her first and middle names in her publications. "I was the first woman the university's English department had ever hired into a tenure-track position, and so I was careful to publish under the gender-neutral name of N. J. C. Andreasen," she recalled in an article she wrote in *The Atlantic* many years later. The pressure on women who wanted to go into medicine was just as intense. Women did not have much presence in postgraduate education and were generally not welcome in medical schools. And now here she was, sitting beside her former students in virtually all-male premed classes, aspiring to become a doctor.

Despite the barriers, Andreasen succeeded at her goal. By the 1980s she had become a world expert at PET (positron-emission tomography), a method in which a radioactive substance is injected into a part of the body—in the case of Andreasen's studies, the brain—in order to create a picture of the tissues. From the point of view of neuropsychiatry and the new field of neuroscience, PET scans were the first giant technological leap beyond Berger's EEG.

The technology of PET scans today is very different from what it was then. "This was before the imaging boom of the nineties," Andreasen tells me. "Back then, you needed to work with a radio-chemist, someone who knows physics, and a physician; you had to know brain anatomy and statistics very well; and you had to be comfortable working with programmers. It wasn't like today, where everything comes in software packages that you just download. Today, you even get the statistics and the brain anatomy laid out for you."

Andreasen's hard work paid off: She would rediscover the peculiar patterns of electrical energy produced by the idling brain—the same energy that Berger had documented and Raichle would later study. Though Raichle would eventually coin the term *default mode,* Andreasen called that manner of the brain's operation the REST state. The acronym stands for "random episodic silent thinking," but it is also tongue-in-cheek, because her point was that when a person's mind seems to be at rest, it isn't. It is just processing information unconsciously, in a different manner.

To understand how Andreasen made her discovery, you have to know a bit about how brain-imaging experiments are done. Like all scientific experiments, an imaging experiment involves a control task. The idea is for the researcher to subtract the activity readings generated in each part of the brain during the control task from the readings generated by the same brain region during the experimental task.

In many experiments, the control task was simply to lie still. In that case, researchers typically gave their subjects instructions like "Let your mind go empty." They thought there would be little going on in the brain during that kind of relaxed state. "That assumption bothered me," says Andreasen. "I doubted that a mind can ever be 'empty.'" So she decided to analyze the resting brain activity itself, rather than use it as a baseline in some other study.

That was when Andreasen came to the same realization that Berger had come to decades earlier. "At rest, there wasn't just a little bit of activity—there was a great deal, and it was concentrated in certain structures," she says. That was a shocking contradiction to the conventional wisdom—just as shocking as it was in Berger's day. But what really fascinated Andreasen was *where* the brain activity was located. It took place in a network consisting of several structures that were previously thought to have little to do with one another—now called the default network.

Even more intriguing, Andreasen says, "it wasn't just a din—it was a symphony. The activity within the structures varied from second to second as it always does, but the different areas, though not all adjacent, were firing in synchrony." The synchronous firing of the three diverse regions told Andreasen that she was on to something.

A lot is written about the size of the human brain, and especially about the size of our prefrontal cortex. But scientists are now beginning to believe that might be a red herring, and that perhaps even more important to our intelligence, and our psyche, is our brain's degree of *connectivity*.

As I described in chapter 4, the brain is hierarchical, and the Human Connectome Project, launched in 2009, is now working on creating a map of the neural connections between structures on each successively larger scale. Even in 1995, however, Andreasen knew that many functions of the brain are served by coalitions of structures, and that the role of each structure can vary depending upon which coalition it is serving. That the regions were firing in unison meant that she had discovered one of those coalitions.

But what was its significance? Andreasen had remade Berger's discovery, and, with the more sophisticated technology available to her, she was able to learn much more than he could about the network of brain structures involved and the type of activity taking place. However, Andreasen had just scratched the surface, and it wasn't until Raichle performed his more extensive research a few years later that the default network moved to center stage in the world of neuroscience research.

In the past decade, scientists have discovered additional structures that contribute to the default network, and we are still working to better understand its role in the brain. But we do know that the default network governs our interior mental life—the dialogue we have with ourselves, both consciously and subconsciously. Kicking into gear when we turn away from the barrage of sensory input produced by the outside world, it looks toward our inner selves. When that happens, the neural networks of our elastic thought can rummage around the huge database of knowledge and memories and feelings that is stored in the brain, combining concepts that we normally would not associate and noting connections that we normally would not recognize. That's why resting, daydreaming, and other quiet activities such as taking a walk can be powerful ways to generate ideas.

Smarts by Association

The power of the default mode stems from its place of origin in the brain—the components of the default network are all within sub-regions of the brain called *association cortices*. We have an association cortex for each of our five sensory systems and for each motor region, and we have what are called "higher-order" association areas for complex mental processes not associated with movement or the senses. I said in chapter 4 that neural networks that represent ideas can activate one another, creating associations. The association cortices are where those connections are made.

Associations help to confer meaning on what you are seeing, hearing, tasting, smelling, and touching. For example, a brain region called the primary visual cortex detects the basic features of the visual world, such as edges, light and dark, location, etc. But that is just data. What does the data mean? What persons, places, and things are you looking at, and what is their significance? It is an association cortex that defines the objects you identify.

When you read a sign that says NO TRESPASSING, the printed letters create an image on your retina. That is just a reproduction of the lines that make up the letters. The sign's message achieves meaning only when that information is passed on from the retina to the visual cortex to an association cortex that identifies the sign, and the letters and words written on it. And that's just the beginning. The image is then passed on to other association regions where connotation, emotional tone, and your personal memories and experience give the words additional meaning.

No one has firsthand knowledge of how other animals think, but scientists who observe them note that they seem to have little power to make abstract associations. Scientists can show, through elaborate experiments involving concrete objects, that rhesus monkeys can add one and one to get two. But to associate the abstraction of the moon's "orbit" with an ellipse seems to be beyond them. In humans, however, about three-quarters of our cerebral neurons reside in the association cortices; this, as a proportion of the brain, is far more than in any other animal.

Our association neurons are what allow us to think and have ideas, rather than merely react. They are the source of our attitudes, differentiating us from one another and helping to define our identities as individuals. They are also the source of our inventiveness. Our culture tends to view discovery and innovation as materializing out of nothing, the product of the ethereal magic of a gifted intellect. But breakthrough ideas, like mundane ideas, often arise from the association and recombination of what is already lying around in the corners of our minds.

That brings us back to the default mode. "When your mind is at rest, what it is really doing is bouncing thoughts back and forth," Andreasen says. "Your association cortices are always running in the background, but when you are not focused on some task—for example, when you are doing something mindless, like driving—that's when your mind is *most* free to roam. That's why that is when you most actively create new ideas."

As is often the case in neuroscience, one way to better understand the role of a structure or network in the brain is to study the behavior of people in which it is disrupted. Consider the famous case of Patient J, who, due to a stroke in her frontal lobe, lost the functioning of her default mode, and then rather miraculously recovered.

Immediately after her stroke, Patient J lay in bed quietly and remained alert. She responded to requests and instructions, and spoke in answer to speech. But she did not initiate any conversation. In the absence of the internal mental dialogue that produces associations, nothing came to mind.

Think about a typical conversation. If her doctor asked, "How do you like the hospital food?" Patient J might have answered, "It's not very good." A healthy individual might have followed up with something beyond the literal answer. She might have added, "If I wasn't in the hospital already, food like that could have landed me here." Or "But it beats the mystery meat at my child's school cafeteria." But one could think of making such remarks only after retrieving private mental associations such as *bad food* and *food poisoning,* or *hospital food* and *school food.* Answers like that don't come from the immediate surroundings or circumstances. They are expressions of your personality that require you to turn inward. Such thoughts were beyond Patient J. She had lost the ability to generate new ideas, so she lost the ability to converse. After Patient J recovered, she was asked about why she had never said anything except in response to questions. She replied that she didn't speak because she "had nothing to say." Her mind, she said, had been "empty."

The Importance of Being Aimless

I had the pleasure of spending a few years working with Stephen Hawking. For the past five decades or so, Stephen has been living with ALS, a disease that attacks the neurons that control the voluntary muscles. Because he has little capacity for movement, Stephen communicates by choosing words from a computer screen through the clicks of a mouse. It's a tedious process. At first the screen displays a cursor moving from letter to letter. Once he has selected a letter, through another click he can either choose from a list of suggested words that begin with that letter or he can repeat the process to choose the second letter of the word he has in mind—and so on, until he has chosen or spelled out the word.

When we first began collaborating, he accomplished the mouse clicks by employing his thumb. Later, as the disease progressed, his glasses were fitted with a motion sensor so that he could click the mouse by twitching a muscle on his right cheek. If you have ever seen Stephen interviewed on television, the quickness with which he responds to the questions is an illusion. He receives the queries long in advance and requires days or weeks to fashion his answers. Then, when the interviewer asks the question, Stephen simply clicks his mouse to initiate the reading of his response, or the sound editor later adds it.

When I worked with Stephen, he could compose his sentences at a rate of only about six words a minute. As a result, I would typically have to wait several minutes for him to make even a simple response to something I said. At first I would sit impatiently, daydreaming on and off as I waited for him to finish his composition. But then one day I was looking over his shoulder at his computer screen, where the sentence he was constructing was visible, and I started thinking about his evolving reply. By the time he had completed it, I had had several minutes to ponder the ideas he expressed.

That incident led me to a revelation. In normal conversations, we are expected to reply to each other within seconds, and as a result, our volleys of speech come almost automatically, from a superficial place in our minds. In my conversations with Stephen, the stretching of those seconds to minutes had a hugely beneficial effect. It allowed me to more profoundly consider his remarks, and it enabled my own ideas, and my reaction to his, to percolate as they never can in ordinary conversations. As a result, the slowed pace endowed my exchanges with a depth of thought not possible in the rush of normal communication.

That rush doesn't affect only in-person conversations. We rush to answer texts, pound out emails, flit from link to link online. We have more assistance from automation and technology than ever before, but we are also busier than ever before. We're bombarded with information, with decisions to make, with tasks on our to-do list, with the demands of work. Adults today typically access their smartphones for an average of thirty-four short (thirty seconds or less) sessions daily, not to mention longer periods for phone calls, to play games, etc. Fifty-eight percent of adults check their phone at least every hour, while eighteen- to twenty-four-year-olds exchange, on average, 110 text messages each day.

The impact of technology can be positive. We are more connected to our friends and family. We have easy and almost constant access, over cell phone or tablet, to television shows, news websites, games, and other apps. But we are also expected to be available always, and everywhere, and because we can work at home and are more connected to our employer, we may be expected to work or be on call virtually all the time. Even our connection to friends and family has a downside, for it can be addictive.

In one study in which participants were asked to refrain from texting for two days, they reported feeling "annoyed," "anxious," and "agitated" when unable to text those they were close to. In another study, iPhone users were found to suffer from anxiety and an increase in heart rate and blood pressure when they were prevented from answering their ringing phone. Another showed that 73 percent of smartphone users feel panicked if they misplace their phone. And yet another documented that many people can't help being on their phone, even when they know they shouldn't. These are classic signs of addiction, and such syndromes are becoming so serious, and so common, that psychiatrists have begun to fashion names for them, such as iPhone separation, nomophobia (for no-mobile-phone-phobia), or, more generally, iDisorders.

Addiction occurs because the constant bombardment of activity that we've become accustomed to can change the function of our brains. The mechanism is much like that of chemical addiction. The fact that we don't know what we'll come across when we check our favorite social media site or our email produces anticipation in our brain, and when we find something of interest, we get a little surge in our reward circuitry. After a while, you become conditioned to the rush, and you get bored in its absence. Meanwhile, beeps, swishes, and harp notes continually remind us that a reward could be waiting.

Remind you of the one-armed bandits in Las Vegas? Says David Greenfield, psychiatrist and founder of the Center for Internet and Technology Addiction, "The Internet is the world's largest slot machine, and the smartphone is the world's smallest slot machine." Video games, including those simple ones you can play on your phone, are even worse. To quote one study, "A massive increase in the amount of dopamine released in the brain was indeed observed during video game play, in particular in areas thought to control reward and learning. The level of increase was remarkable, being comparable to that observed when amphetamines are injected intravenously."

The result of our addiction to constant activity is a dearth of idle time and, hence, a dearth of time in which the brain is in its default mode. And though some may consider "doing nothing" unproductive, a lack of downtime is bad for our well-being, because idle time allows our default network to make sense of what we've recently experienced or learned. It allows our integrative thinking processes to reconcile diverse ideas without censorship from the executive brain. It allows us to mull over our desires and shuffle through our unattained goals.

Those internal conversations feed our ongoing first-person narrative of life, and they help develop and reinforce our sense of self. They also allow us to connect divergent information to form new associations, and to step back from our issues and problems to change the way we frame them, or to generate new ideas. That gives our bottom-up elastic thinking networks the opportunity to search for creative, unexpected solutions to tough problems. On that night she invented her character Frankenstein, had Mary Shelley possessed a cell phone, rather than resting and letting her thoughts wander, she might have reached for the device; its many lures might have attracted her conscious attention and suppressed the emergence of her idea.

The associative processes of elastic thinking do not thrive when the conscious mind is in a focused state. A relaxed mind explores novel ideas; an occupied mind searches for the most familiar ideas, which are usually the least interesting. Unfortunately, as our default networks are sidelined more and more, we have less unfocused time for our extended internal dialogue to proceed. As a result, we have diminished opportunity to string together those random associations that lead to new ideas and realizations.

It's ironic, but the technological advancement that makes elastic thinking ever more essential also makes it less likely that we'll engage in it. And so, if we are to exercise the elastic thinking that is demanded by our fast-paced times, we have to fight the constant intrusions and find islands of time during which we can unplug. In the past few years, that issue has become so urgent that a relatively new field called eco-psychology has suddenly flourished.

Ecopsychologists are gathering scientific evidence to back up their assertions, but many of their recommendations aren't new. For example, they suggest that one way to set aside dedicated quiet time is to disconnect and take refuge in activities like jogging or taking a shower. Walks are also useful—but you have to leave your cell phone at home. Such walks allow your default mode to kick in, and they help restore your top-down executive functions. When you get back to join the rat race, you'll be revitalized—but only if you walk in a quiet area. Noisy urban neighborhoods are filled with stimuli that capture and direct your attention—for example, when you need to avoid bumping into someone or getting hit by a speeding car. But if the act of walking or running can free your mind, so can taking a few minutes in the morning after you wake up to simply lie in bed. Don't think about your schedule that day or ponder your to-do list but, rather, take advantage of your quiet state to stare at the ceiling, enjoy the comfort of your bed, and relax a little before popping up to face the world.

At work, rather than always pushing on when you are stuck and unable to resolve some complex issue, you can schedule a bout of mindless tasks as a break. Keeping something as trivial as a shopping list in your mind can impede elastic thinking, so try to purge your thoughts of what you had been working on, and of the things you still have to do. If you successfully clear your brain, you'll be getting some easy work done while simultaneously freeing your elastic mind to look for a breakthrough solution to your impasse. Even an hourly pause to walk to the water cooler can help. The interludes will give your elastic mind a chance to process—and question—what that last hour of concentrated thought produced.

Surprisingly, procrastination can help, too. Research shows a positive correlation between procrastination and creativity, because by putting off conscious attempts to solve problems and make decisions, we provide ourselves more time to fit in those episodes of unconscious consideration.

Leonardo da Vinci had such great respect for unconscious processing that while he was working on *The Last Supper*, he would sometimes suddenly quit for a while. The clergyman who was paying

Leonardo did not appreciate those hiatuses. As art historian Giorgio Vasari put it, "The prior of the church entreated Leonardo with tiresome persistence to complete the work, since it seemed strange to him to see how Leonardo sometimes passed half a day at a time lost in thought, and he would have preferred Leonardo, just like the laborers hoeing in the garden, never to have laid down his brush." But Leonardo "talked to him extensively about art and persuaded him that the greatest geniuses sometimes accomplish more when they work less." The next time you stare out the window, remember, you aren't slacking off—you're giving your artistic side a chance to do its work. And if you don't tend to take such breaks, you might find it beneficial to make room for them.

7

The Origin of Insight

When the Unimaginable Becomes the Self-Evident

On December 21, 1941, two grim weeks after Pearl Harbor, President Franklin Roosevelt told his Joint Chiefs of Staff in a meeting at the White House that it was imperative that Japan be bombed as soon as possible, both to boost morale at home and to plant the seeds of doubt in the Japanese people, whose leaders had told them they were invulnerable. Despite the urgency of the mission, it seemed like an impossible task: No bomber had anywhere near the range necessary to fly to Japan.

One cold day a few weeks later, a submarine captain named Francis Low was reminded of Roosevelt's challenge as he watched bombers on practice runs at a naval airfield in Norfolk, Virginia. The rectangular outline of an aircraft carrier deck had been painted on a runway to provide the bombers their mock target. Like everyone else who'd been told about the challenge, Low had been drawing blanks. A lifelong navy man and a submarine captain by training, bombers were far from his area of expertise. But as he watched the shadows of the planes cross that painted outline, an idea suddenly exploded into his consciousness. It was an idea an expert would have dismissed as absurd. What if they launched their bombers from the deck of a carrier?

It was an instance in which the key to solving the problem was ignorance, or at least pretending that what you know isn't true. Low wasn't completely ignorant—he understood some of the many reasons his idea "couldn't work"—but he decided to ignore them. He instead embraced the assumption that it *had* to work, and began analyzing how to overcome the obstacles.

There were many such obstacles. Aircraft carriers are designed to transport nimble and lightweight fighters, not bombers, which are too heavy to take off from a carrier's limited runway. Bombers are also not very maneuverable and, hence, are easy to shoot down, so they must be escorted by fighters, but an aircraft carrier hasn't the space to carry both. Most important, even if a bomber could somehow be placed on a carrier and the carrier could take it close enough to Japan for it to be within bombing range, the high tail and weak tail structure of a bomber made it impossible to install a landing hook, and so the returning bombers would not be able to land on the carrier. Low didn't have most of the answers, but he didn't accept the prospect that none could be found.

After returning to Washington, he went to see his commanding officer, Admiral Ernest King. Low was always uncomfortable in his superior's presence, and now he was especially nervous. His suggestion was bound to strike the stern admiral as outlandish. Low waited until King was alone and then, during a pause in their conversation, he blurted out his idea.

Though the scheme seemed unlikely to succeed, the times were desperate. So over the next months, bombers were stripped down to their bare essentials to reduce their weight and fitted with extra fuel tanks to extend their range. Pilots were trained to take off from a short carrier deck and to fly low enough to evade Japanese radar, eliminating the need for a fighter escort. And the problem of not being able to land on the carrier was "solved" by accepting the notion that, after dropping their bombs, the pilots could fly onward to land or ditch their planes in the Chinese or Soviet countryside. Those countries had denied the Americans permission to use their territories to stage and launch an attack, but if the planes were merely to land there, they would not even have to know about it. The crews of the ditched planes, unfortunately, would be left with the huge challenge of finding their own way to Allied lines.

Army Air Forces Chief of Staff General Henry H. Arnold assigned the technically astute Colonel Jimmy Doolittle to organize and lead the raid, and Doolittle gathered eighty volunteers to man sixteen B-25s for the mission. Because the idea that American bombers could reach Japan seemed so implausible, they encountered almost no anti-aircraft fire; in fact, many Japanese on the ground waved to them, thinking they were part of a practice mission conducted by their own air force. All told, the planes dropped sixteen tons of bombs on Japan, mostly on the Tokyo area. After the raid, the crews of all the planes either crash-landed or bailed out in provincial China; all but six of the crewmen survived.

To grasp how absurd Low's idea must have seemed to the experts of the day, consider this: The Japanese, desperate to eliminate the risk of any more such raids, found it unimaginable that the planes had been carrier-based. They became convinced that the attack had come from Midway Atoll, the only possible land source—and they sent their fleet there to take that island. The U.S. Navy, one step ahead, was lying in wait and sank all but one of their carriers. The Japanese fleet was essentially crippled, a setback that military historian John Keegan called "the most stunning and decisive blow in the history of naval warfare."

Sometimes the most powerful revelation one can have is that circumstances have changed. That the rules you are accustomed to no longer apply. That the successful tactics may be tactics that would have been rejected under the old rules. That can be liberating. It can spur you to question your assumptions and help you rise above your fixed paradigms and restructure your thinking.

In this case, an aircraft carrier was a piece in a standard war-planning puzzle. So, too, was a bomber. Under normal circumstances, those pieces didn't fit together. It took Low to recognize that in the aftermath of Pearl Harbor, the puzzle had changed. A course of action that clearly didn't fit the game of conventional warfare *was* appropriate to the demands of that moment in history. History—and ordinary human life—is full of opportunities missed by not recognizing that change has occurred, and that the previously unthinkable is now doable.

When asked how he had come to his plan, Low said it was "fortuitous," as if he'd walked into a Chinese restaurant and found his idea written in a fortune cookie. No doubt that's how his conscious mind perceived it. But today we know that insights like Low's are not accidental. They are the result of a complex process that your unconscious brain engages in after your conscious logical reasoning, constrained by commonly accepted rules and conventions, fails you.

In the last chapter we learned about the default network and saw that our brains are at work making associations even when we are not consciously focused on anything in particular. Most of those associations never reach our conscious awareness. In Low's case, desperation pushed his mind to consider a last resort, an idea that otherwise would have been rejected. Here we will explore the mechanism that brings those associations into your consciousness, and what determines whether the ideas presented to your awareness are mere conventional notions or brilliantly original insights.

Splitting the Brain

Roger Sperry pondered what he had found. It was the late 1950s. He had been experimenting on animals on which he had previously operated to sever the corpus callosum, the structure that sits between the brain's right and left halves. He was aware that most of his colleagues thought his work was a waste of time—that the corpus callosum played an uninteresting mechanical role in the brain. They considered it a kind of girdle that kept the hemispheres "from sagging." Sperry envisioned a grander function: allowing the right and left hemispheres to communicate. And now, with his discovery, he had astonished even himself.

The conventional wisdom was that communication between the hemispheres was largely unnecessary. The left side of the brain was seen as being responsible for cerebral functions ranging from understanding language to arithmetic reasoning to controlling willful movement. The right hemisphere, by contrast, was said to be generally lacking in higher cognitive function—mute and unable to talk or write, even in left-handed individuals. As a result, doctors commonly told patients who had had a stroke damaging the right side of their brain that they were lucky, because that hemisphere "doesn't do much." Some of Sperry's colleagues even referred to the right side of the brain as being "relatively retarded." Given that, why would communication between the hemispheres be important?

Sperry wasn't buying the conventional wisdom. It was based on the observation of patients with brain lesions, and he didn't trust that research, because scientists had little control over it—patients never walked in with damage to exactly the brain structure one wished to study. Sperry reasoned that he could do better. Employing his excellent skills as a laboratory surgeon, he could excise precise regions of a brain and observe changes to behavior due to that exact deficit—though only in animals, of course. And that's exactly what he did to investigate the role of the corpus callosum: He surgically severed that structure. An animal that's undergone this procedure is said to have a "split brain."

At first Sperry had been disappointed. He found, just as everyone warned him would be the case, that the surgery had little effect on the animal's day-to-day behavior. But then he designed a new battery of tests to isolate the individual hemispheres from each other more carefully. It was the results of those experiments that now astonished him.

In one experiment, Sperry covered one eye of split-brained cats. As they viewed the world through the uncovered eye, he taught them to discriminate between a triangle and a square. Geometry is not a cat's best subject, but they eventually became good at the task. Then Sperry switched the covering to the other eye and again tested the cats. Did the opposite hemisphere benefit from the training? Sperry found that with the corpus callosum cut, cats were "unable to perform with one eye a visual-pattern discrimination learned with the other." By severing the corpus callosum, he had precluded the communication between hemispheres.

Through a series of careful experiments such as that, Sperry learned that he could interact independently with each hemisphere of an animal's brain, and found them to be startlingly self-sufficient when it came to processing information. "Each of the divided hemispheres," he would later write, "has its independent mental sphere or cognitive system—its own independent perceptual, learning, memory, and other mental processes. It is as if each hemisphere is unaware of what is experienced in the other."

In other words, Sperry claimed, the animals seemed to have two minds within them. Their individual capabilities and potential for independent thought are not normally apparent, for in a healthy individual the minds are highly connected through the corpus callosum and work in harmony. But when you sever that connection, they reveal their individual selves.

Sperry viewed his results as revolutionary, but others didn't see them that way. They believed that his conclusions were valid only for "lower animals." Sperry knew that his challenge was to prove that it worked the same in humans, but how? He couldn't exactly haul a human into his lab and sever the corpus callosum the way he did with animals.

That's when Joseph Bogen, a neurosurgeon, showed Sperry an essay on epilepsy he had written, called "A Rationale for Splitting the Human Brain." In the 1940s, surgeons had experimented with severing the corpus callosum in order to reduce the fury of seizures in patients with severe epilepsy, and Bogen was thinking about resurrecting that approach. He suggested that if he were to perform any of those surgeries, perhaps Sperry would like to study his patients. It was just the break Sperry needed.

In 1962, Bogen operated on the first of a series of sixteen patients he would invite Sperry to test. The results of those studies confirmed what Sperry had found in his animal studies: that the conventional wisdom about the roles of the hemispheres was wrong. The two hemispheres of our brains *are* something like independent beings. For example, in one case a patient was asked how many seizures she'd had recently. Her right hand shot up and showed two fingers, indicating she'd had that number of seizures. Her left hand, controlled by the opposite cerebral hemisphere, pulled the right hand down. Then her left hand rose and indicated just one. Her right hand went up again and the two hands tangled, fighting each other like angry children. Eventually the patient said that her maverick left hand often "did things on its own."

By naming the left hand as the maverick, the patient seemed to be taking the side of her right. That's because the right hand is controlled by the brain's left hemisphere, which also controls speech. This illustrates an important point: Though the right cerebral hemisphere is not, as prevailing thought put it, "retarded," the two hemispheres do have some different capabilities. For example, though the right hemisphere can comprehend spoken words, it cannot speak them. So when a split-brain patient talks, it is the left hemisphere that is expressing itself.

Understanding those hemispheric differences would later prove to be the key to understanding the origin of ideas like the sudden insight experienced by Francis Low. But in Sperry's day, scientists were skeptical, and as word spread, the new research quickly became controversial. After all, it not only challenged long-held scientific ideas—it could also threaten philosophical or even theological beliefs. Is the notion that each of us is an "I" an illusion? If there are two "beings" within us, does that mean we are each two people or have two souls? Those aren't questions that science concerns itself with, but Bogen didn't want to subject himself to possibly devastating attacks by scientists or anyone else. He asked to withdraw his name from Sperry's publications. The work, however, stood the test of time, and Sperry received the Nobel Prize for it in 1981.

Sperry died in 1994—about three decades after his groundbreaking research. During those years, scientists continued to explore the roles of the two hemispheres, but progress was slow. Sadly, it was shortly after Sperry died that the pace picked up because of the availability of fMRI and other new brain-imaging technologies.

The past two decades have seen an explosion in understanding regarding the roles of the two hemispheres and the structures within them. And in recent years, one startling conclusion of that research is that a certain structure in the once derided right hemisphere has a special talent for generating the original ideas that are in great demand when an organism is confronted with novelty or change, or a seemingly intractable intellectual challenge.

The Connection Between Language and Problem-Solving

The origin of new ideas is one of the concerns of the field of cognitive psychology, the study of how we human beings think. Until relatively recently, scientists could draw their conclusions only from the indirect evidence provided by behavioral studies—and through guesswork. But during the 1990s, the field gave birth to a new kind of science called cognitive neuroscience, which employed evidence derived from the new brain-imaging technologies. The field's pioneers aimed to use those instruments to study the physical processes in the brain that produce our thoughts, feelings, and behavior—and to understand how they are related to each other, and how we can manipulate them. The new technologies, they recognized, offered the power not only to understand the way we think but to help us change it.

One of those pioneers was John Kounios, then a young assistant professor at Tufts University. Kounios focused on employing a technology to study what are called ERPs to investigate how the brain processes language. The acronym stands for "event-related potentials," the electrical activities in the brain that result from an internal or external stimulus. It had been known since Berger's work that you can measure ERPs employing Berger's EEG, but the new technology coupled that with powerful computers to create a far more precise picture.

One day, while analyzing the timing of neural activity as the brain strives to understand the meanings of words and sentences, Kounios's own brain made a new association. He suddenly saw an analogy between the processes involved in understanding the meaning of a sentence and the elastic thinking required to forge a response to a daunting mental challenge—the kind of thinking that had led Low to his grand idea during World War II, or that you might employ to react to a new circumstance in life or to solve a riddle or an intellectual puzzle.

How are sentences like puzzles? Each sentence is an ordered list of words and punctuation. But most words have multiple meanings, and those meanings can combine in different ways, depending upon the grammar and context. That's the puzzle: to choose among the various definitions of the individual words in such a way that the whole sentence has meaning, a meaning that fits in the larger context if there is one. It's really an exercise in integrative thinking: Rather than try to decide the meaning of each word as it is spoken, your brain keeps the whole sentence and the larger context in mind while making sense of the words.

To accomplish that, as we hear or read each word, we hold its possible definitions in our working memory while our brain processes the other words of the sentence and considers their range of definitions. Only at the end do we put it all together. For example, consider the sentence "The cooking teacher said the young children made bad snacks." When you read that, your unconscious mind quickly sorted through the various meanings of all the words and chose the appropriate ones. Now read this sentence: "The cannibal said the young children made bad snacks." Chances are, this time you assigned a different meaning to the word *made.* This sentence differs from the prior one by only one word, but that word alters the context and, hence, your brain's interpretation of the words that follow. Similarly, as a bestselling book on the importance of grammar pointed out a few years ago, when you read the phrase "eats shoots and leaves," you understand one meaning for the words *shoots* and *leaves,* but if commas are added so that you read "eats, shoots, and leaves," then you assign the words a different meaning.

One of the amazing things about the human brain is that when we hear or read sentences, the appropriate meanings come to mind quickly and without conscious effort. But that's only because we have a brain with an unconscious tuned to doing that—thanks to the millions of years of evolution that provided our cerebral hardware and the many thousands of hours of exposure to our native language that allowed us to program it. You get an appreciation of how great a gift that is anytime you listen to or read a language you don't know well. That task is slow and effortful, because your unconscious hardware is not yet trained and you must consciously puzzle out word meanings.

In the 1950s, when the digital computer was still a new invention and information scientists thought that artificial intelligence would soon rival that of humans, computer linguists vastly underestimated the power of our unconscious language processing. They thought it would be easy to program a computer to replicate it. Their lack of success is illustrated by the story of an early computer that translated the homily "The spirit is willing, but the flesh is weak" into Russian and then back into English, and obtained "The vodka is strong, but the meat is rotten." Before it was converted to the neural-net approach, Google Translate still made analogous mistakes.

The Trial of the Hemispheres

When Kounios became curious about how the brain's language comprehension ability might relate to other kinds of problem-solving, he began scanning the existing literature. There he found a lot of work in which psychologists had arranged to show various intellectual challenges to just one side of the brain or just the other, as Sperry had done in his cat experiments.

These scientists had uncovered intriguing hints that the right hemisphere played a special role in generating imaginative ideas, but those papers relied on the self-reports of their subjects to determine what they were thinking. Unfortunately, many people's self-knowledge doesn't extend much beyond recognizing when they want a beer. As a result, even when biases are not in play, self-reporting can be unreliable.

How unreliable? The habit of drawing conclusions without really knowing why we make them bothered me when I worked, for a season, as a writer on the staff of *Star Trek: The Next Generation.* Unlike my personal choices, or those I made in my physics research, the choices we made on that television series could have a big effect on people—for example, when we bought or rejected script submissions or made judgments about casting. And so, those times I participated in casting decisions, I would always ask what the producer(s) saw in the chosen actor or actress. They'd say things like "He has a presence." To that, my literal, analytic mind would ask, *What does that mean?* Who doesn't have a presence? Only someone who didn't show up for the audition, right? In hindsight, I realize that the descriptors the producers gave portrayed a connection they *felt,* on the unconscious level. The source of the connection, however, was usually difficult for them to articulate.

Based on what scientists have discovered in the decades since, we now know that the architecture of your brain puts out of reach the behind-the-scenes influence that your unconscious mind exerts on your thinking. As a result, though introspection can help illuminate aspects of *conscious* reasoning, of analytical problem-solving, it cannot provide much insight into elastic thinking. Yet it is those subliminal elastic processes that Kounios suspected lead to the moments of sudden insight that are so famous in the annals of discovery and innovation, and in the revelatory moments of our lives. As a result, despite dozens of behavioral studies, as long as the research relied on self-reports, the science of insight hardly progressed.

Before we get too far into the science of insight, it is helpful to take a moment to consider what cognitive psychologists mean by "idea" and "insight." In normal parlance, an idea can be a composite, developed over a long period of time and made of many component notions, as in "the idea of the quantum." In the science of thinking, however, an "idea" usually refers to something simpler, the complexity of which can be contained in a single thought, and that pops suddenly into our consciousness. An "insight" is defined as an idea (of that sort) that represents an original and fruitful way of understanding an issue or approaching a problem.

"The origin of insight was a fascinating puzzle," says Kounios, "and I knew that solving it could be important for people's economic success. Yet for some reason there were virtually no neuroscience studies on it back then. That was good because my lab was a small one, and the big, well-funded labs have a huge advantage. They had better equipment and more people, and could crank out work very fast. But they weren't working on insight." And so Kounios made the fateful decision that for the next phase of his career he would work to understand those moments employing the technological tools he had used to study the neural activity of decoding sentences.

At the same time that John Kounios was starting to focus on the physiological basis for insight, a few hundred miles away at the National Institutes of Health, so was Mark Beeman. Like Kounios, Beeman studied language processing. And like Kounios, he had read about Sperry's pioneering work in college and was surprised at how many people continued to ignore the role of the brain's right hemisphere.

Like the skeptics in Sperry's day, their lack of interest was based on the observation of stroke patients and others with right-hemisphere brain injuries. Those patients' mental deficits were often more subtle than the deficits of the patients with left-hemisphere damage, but Beeman was convinced that they were significant. For example, people with certain left-hemisphere damage lose their speech capability, while those with right-hemisphere damage don't. But people with right-hemisphere damage do have some language issues. Although they can still speak, "they have trouble understanding jokes and metaphors, and finding the theme of a story, or drawing inferences," says Beeman. To Beeman, those issues were the key to understanding the right hemisphere's role.

What did those language deficits have in common? What does getting a joke have to do with understanding a metaphor? Like Kounios, Beeman thought about how our brains puzzle out language. After you encounter a word and your unconscious pulls up all of its various possible meanings, it determines the probability that each meaning might be appropriate to the sentence under consideration. The most obvious and common meanings start with the highest probabilities. As you hear more of a sentence, those probabilities are updated according to the context.

The associations you attach to the meanings of words play an important role in that process. As you hear a sentence, your brain looks for where the associations of all the words in the sentence overlap, and it uses that information to make its best guess regarding what the speaker is trying to get across. For example, in the case of the sentence "The cooking teacher said the young children made bad snacks," the context associated with "cooking teacher" tells your brain that the appropriate meaning for the phrase "made bad snacks" has to do with the creation of food. On the other hand, when you read the sentence "The cannibal said the young children made bad snacks," the context associated with "cannibal" tells you that "made bad snacks" has to do with being eaten as food.

Although those are the most probable and obvious interpretations of each of those sentences, they each could have been construed the other way. The author of the sentence "The cooking teacher said the young children made bad snacks" could conceivably have meant that the cooking teacher had just eaten the kids as a snack, and the author of "The cannibal said the young children made bad snacks" might conceivably have wanted to convey that the cannibal disdained the children's cooking ability. Your unconscious mind noted those possibilities, but it probably didn't make you aware of those unlikely interpretations (psychologists call them "remote").

Before an idea passes to your awareness, your brain conducts a kind of trial in which it considers all the evidence for the various meanings your unconscious mind has produced. Only then does it pass to your consciousness what it has deemed to be its best guess. As the brain weighs the meanings, its two hemispheres slug things out. Your left hemisphere advocates for the obvious and literal meanings, while your right hemisphere takes on the underdogs, the meanings that at first may seem remote, a bit of a stretch, but that are sometimes the correct interpretation.

Beeman realized that when you look at the roles of the hemispheres that way, the language deficits of patients with damage to their right hemisphere make sense. Take metaphors. They are figures of speech in which a word or phrase that usually means one thing is used to mean another. The word *light* usually refers to the electromagnetic phenomenon, but in "the light of my life," it means joy or happiness. The word *heart* usually refers to the organ, but in "broken heart" it denotes an emotional state. When you understand a metaphor, it's because your right brain advocated the kind of fuzzy associations that allow one to understand these expressions—which explains why, if you had a stroke in the language center of your right brain, you wouldn't be able to understand metaphors.

Jokes often rely on a similar process. Consider this, from a monologue by Conan O'Brien: "It's being reported that, because of the birth of his baby daughter, Chris Brown has decided to stop calling women *hoes* in his music. He says he's going to go with the more traditional term, *bitch*."

The term *traditional* normally conjures the context of long-established culture, even ancient or religious practices. By contrast, the use of the word *bitch* in hip-hop circles to refer generically to any woman is relatively recent. So this joke would confuse your left brain—when *traditional* is understood in the usual way, *bitch* is a non sequitur. But your right brain gets it—it allows a broader, fuzzier interpretation of the term *traditional,* one that allows for sarcasm. Beeman was impressed by the "fuzzy logic" capability of the right brain, and curious about whether it might have applications outside of language processing. "And then it struck me," he says. "I realized that the right hemisphere's role in insight is like its role in language." He and Kounios were now both headed in the same direction.

The Lessons of CRAP

Kounios and Beeman finally crossed paths in late 2000. Kounios had conducted his studies of ERPs using the EEG, but Beeman had learned the new fMRI technology. In determining timing, the EEG is far superior, but fMRI provides more precise maps of brain structure and activation—exactly what the EEG isn't good at. "When we thought about that," Kounios told me, "that's when it clicked. We realized that, working together, we could say both when *and* where things happen." They agreed to be partners.

Kounios and Beeman decided to design a set of parallel studies. They would each recruit their own subjects and record their brain responses in their own labs, employing their respective technologies. But they would present the subjects in both labs with the same puzzles. In that way, Kounios could nail down the timing of brain responses, and Beeman the geography. By combining their data, they would obtain a complete picture of which brain structures are activated, as well as how they are orchestrated.

Kounios and Beeman sought to design a word game that could be solved through either unconscious insight *or* conscious analytical reasoning. They decided to use a kind of brainteaser patterned after the riddles in what psychologists call the remote associates test, or RAT. They called their variants "compound remote associate problems," or CRA, shortening the more obvious choice of acronym because, although some research can be described by that word, nobody wants it to appear in their papers.

Here's how their CRAs work. Subjects are shown three words—for example, *pine*, *crab*, and *sauce*. They are asked to think of a "solution word," a fourth word that can form a familiar compound word or phrase when appended to each of these three words. The solution word can come either before or after the given words. For example, consider the word *nut*. "Pine nut" works, as does "nut sauce." But neither "crab nut" nor "nut crab" makes sense, so *nut* cannot serve as the solution word.

It will help you appreciate the mental process Kounios and Beeman monitored in their respective labs if you try to solve the *pine-crab-sauce* CRA yourself. Their subjects solved only 59 percent of the puzzles, so don't worry if you don't succeed. It is getting a feel for the process that is important, so I suggest that you give yourself up to half a minute, and then keep reading. We'll get to the solution shortly.

Kounios and Beeman designed the puzzles so that two of the three words bring up strong and obvious associations. In this case, *pine* seems to imply a type of tree, so words like *cone* (*pinecone*) and *tree* (*pine tree*) come to mind. *Crab* brings to mind the crustacean, so words like *cake* (*crab cake*) and *meat* (*crabmeat*) quickly flash into consciousness. But since none of those words combines well with both of the other words, you come to realize that the solution word probably has nothing to do with trees or crustaceans. In other words, to solve the puzzle, you must let go of your immediate association of *pine* with trees and *crab* with crustaceans, and let weaker, less obvious, remote associations come into play. That's what makes it hard, but that is what insight is all about—making the unusual associations, through elastic thinking, that analytical thinking can discover only with difficulty.

It's possible to attack the puzzle through conscious, analytical thinking. You start, say, with *crab* and generate an "associate" of the word, such as *crab cake*. If, as in this example, your word (*cake*) does not form a word or phrase with *pine* or with *sauce*, you try again, and keep trying until the solution word is hit upon. But that can prove to be a very laborious process. On the other hand, those who use insight allow their minds to relax and wander until they find the answer, an idea that seems to appear suddenly, from nowhere. In this case that solution word is *apple.*

In the Kounios-Beeman experiments, the subjects were given thirty seconds for each trial. Most employed insight in some of the trials and analytical reasoning in others, though, despite the short time allotted, some also switched their approach in the midst of a trial. In each case, the subjects reported which method had led them to the solution. About 40 percent more puzzles were solved using insight than using logical analysis, and it was the thought processes leading to those solutions that Kounios and Beeman sought to understand.

Beeman's subjects solved their CRAs while lying inside an fMRI machine. Kounios's solved theirs while sitting in a stiflingly hot lab in which the air-conditioning was broken, wearing a kind of shower cap that anchored the dozens of electrodes that Kounios attached to the subject's scalp and face. "The subjects' sweat kept interfering with the readings," remembers Kounios. But it was worth it, because the experiment would become a classic. The results they discovered illuminated, as never before, the mental process that produces human insight.

Deconstructing the Insight Process

What Kounios and Beeman discovered surprised everyone. The headline is that, although our conscious experience of the moment of insight is sudden, it comes from a long string of behind-the-scenes events that mirror the processes involved in understanding language, and with a similar division of labor among the right and left hemispheres.

Here is how Kounios and Beeman deconstruct the insight process, whether in word games such as their CRA puzzles or in other domains. When you are presented with a problem, the brain begins to sort through possible solutions, just as it sorts through the possible meanings of a word in a sentence. That occurs quickly, and outside of your consciousness. Your left brain makes the obvious associations and raises all the obvious answers. Your right brain searches for the obscure associations and the oddball answers. To be precise, Kounios and Beeman found that the oddball answers arise from increased neural activity in a fold of brain tissue just above the right ear, called the right anterior superior temporal gyrus (aSTG).

The different approaches taken by the right and left hemispheres of your brain illustrate the wisdom of Sperry's observation, more than half a century earlier, that they are like two independent cognitive systems. Deep within your unconscious mind, each hemisphere battles to have its ideas accepted by the jury of your executive brain and passed to your conscious awareness. But it seems that there is also a judge who can influence the proceedings. It is a mysterious brain structure that neuroscientists call the anterior cingulate cortex, or ACC, which lies just above the corpus callosum.

One role of the ACC is to monitor other brain regions. I call it a judge because, though the science isn't yet settled, scientists theorize that when the right and left brain are taking their different approaches to a problem, the ACC may step in and act to control the relative strength with which the two hemispheres are heard.

When you first consider a problem, your executive brain gives you a narrow focus. It ignores odd ideas and directs your conscious awareness toward the tried and true, the literal, the logical or most obvious of the possible answers your associative brain produces. The left brain's guesses thus tend to come to your awareness first. That makes sense, because ordinary or unoriginal ideas are usually sufficient.

According to the scientists' theory, if those initial ideas don't lead to an answer, your ACC broadens the scope of your attention, loosening your focus on the conventional ideas of your left hemisphere and allowing the more original ones being proposed by the right to surface.

Roughly speaking, your ACC accomplishes that by arranging for the right visual cortex—the part of your right hemisphere responsible for processing visual information—to shut down. This is like closing your eyes so you can concentrate when you are trying to solve a difficult problem, but in this case, the ACC is blocking out only the visual inputs to your right hemisphere. That suppression of visual activity allows the ideas generated in the right aSTG to take up the slack and grow stronger, so that they may burst into your consciousness. That is why the quality of grit is important: When you reach an impasse, you may feel frustrated and be tempted to give up, but that is precisely the moment when, if you keep struggling, your ACC may kick into action and your most original ideas can begin to surface.

Insights are among the greatest accomplishments of our elastic thought process, and to have finally understood the mechanism that takes us from impasse to insight was a great accomplishment. But Kounios and Beeman made another important discovery. When they looked back at their subjects' brain activity, they could see that sometimes there was distinctive neural activation in those who would go on to solve the problem through a sudden insight, patterns that came long before the insight. In fact, they became apparent several seconds *before the problem was even presented.*

The activity apparently reflected an insight mindset. The brains of some subjects appeared to be primed to succeed through insight by their psychological state, which somehow set the conditions in advance for their right hemisphere to be heard. The neural mechanism through which one can take control of that process is not yet understood, but the implication is that you can nurture insight and lay the groundwork for what will later appear to be the spontaneous generation of novel ideas. The key seems to be to approach your problem with a "relaxed" mind, as opposed to focusing intensely on applying straightforward logic.

I experienced that phenomenon when I was a young physicist. I was seeking the answer to a rather complex problem. I had found an unimaginative mathematical approach that I knew would work, but it was involved and tedious. I had been following that approach with great focus for several days, and still had a long way to go, when Friday evening came along. I had previously asked a woman to dinner that night, so I took pains to relax my mind, and I met her at the restaurant. I'd just ordered linguine when, without warning, an elegant trick for solving my problem with relative ease popped into my consciousness. My focus on the straightforward approach had apparently been interfering with my ability to find that superior method.

Now that I'd thought of it, I felt an irresistible urge to work out enough of the mathematical details to confirm that the idea made sense. How do you say to a woman that she is captivating, but could she wait five minutes while you scrawl some equations on your napkin? I'd wanted a romantic evening, but as she put her hand on mine, my mind was stuck on the geometry of an infinite-dimensional space.

I learned the lesson of Kounios and Beeman's work that night: When attacking a tough problem, impatience to make progress can lead one to a suboptimal solution while blocking your ability to find a better one. Adopting a relaxed mindset, on the other hand, can enable original and imaginative answers to emerge. And so, by allowing your mind to loosen up, you can take steps to awaken your ACC and unleash your powers of insight.

For those who are interested, one can practice taking control. Just Google "remote associates test" and take some of the tests you'll find offered online. You can decide, in each instance, whether to focus on an analytical or an elastic approach, and observe the difference in your thinking.

Zen and the Art of Ideas

Kounios tells of the time when a Zen Buddhist meditator visited his lab. He asked the man if he wanted to try a series of CRAs. The meditator agreed. His mind was so focused, however, that the odd word associations demanded by the CRAs didn't come easily. Again and again he failed to produce a solution within the allotted time. He was doing so poorly, Kounios wrote, that he decided to stop the session, to spare the man further embarrassment. But before Kounios did so, the meditator finally got one correct. That was followed by another, and then another. From that point forward, he got nearly all of them right.

The meditator, noting the failure of his approach, had apparently taken control of his state of mind and stimulated his ACC to broaden his perspective. And he maintained that broad attention state in problem after problem to produce, in the end, an outstanding performance.

To appreciate just how impressive that feat was, consider that, over the years, having tested hundreds of subjects on CRAs, Kounios had never seen any evidence that the practice achieved during a single session can improve a subject's results. Only this meditator, with his enormous awareness of his own thought processes and his great ability to control his mental state, realized the importance of mindset, and was able to turn a switch in his brain to bring success.

In recent years, neuroscientists have found that the "Mindfulness of Thoughts" exercise I presented in chapter 4—actually a kind of meditation—promotes the abilities the Zen meditator displayed. A 2012 study, for example, showed that such meditation enhances your ability to broaden your focus at will so that your mind can quickly and freely jump from one idea to another, encompassing both the ordinary and the unconventional.

It's no accident that the trait of mindfulness has again surfaced in our discussion. In chapter 4 I talked about it in the context of freeing yourself from automated thinking. The challenge of insight is the analogous issue of freeing yourself from narrow, conventional thinking. If you are interested in assessing your own degree of mindfulness, you can answer the following questions that research psychologists employ to measure that trait. Simply use this 1-to-6 scale to rate how frequently or infrequently you have each of the everyday experiences listed in the questionnaire below:

1 = almost always
2 = very frequently
3 = somewhat frequently
4 = somewhat infrequently
5 = very infrequently
6 = almost never

Here are the statements:

1. ___ I break or spill things because of carelessness, not paying attention, or thinking of something else.
2. ___ I tend to walk quickly to get where I am going without paying attention to what I experience along the way.
3. ___ I tend not to notice feelings of physical tension or discomfort until they really grab my attention.
4. ___ I forget a person's name almost as soon as I've been told it for the first time.
5. ___ I find myself listening to someone with one ear, doing something else at the same time.

Total: ___

The possible scores on this questionnaire range from 5 to 30. The average score is about 15, and approximately two-thirds of all who take the test score in the 12–18 range.

16% fall here	68% fall here	16% fall here
5 12 15 18		30
relatively mindless	average	relatively mindful

Distribution of mindfulness scores

Increasing your degree of mindfulness is a good way to encourage insight, but it isn't the only way. You can also cultivate insight by adjusting your external conditions. For example, research shows that sitting in a darkened room, or closing your eyes, can widen your perspective; so can expansive surroundings, even high ceilings. Low ceilings, narrow corridors, and windowless offices have the opposite effect. And a well-lit room can make it difficult to ignore objects in your surroundings that stimulate mundane thoughts, shoving aside the imaginative musings of your right hemisphere.

Being able to think without any kind of time pressure is also beneficial for generating insight, because if you have to start on something else soon, your awareness of that can pull your mind back to the external world and block an unconscious idea from popping into your consciousness. Perhaps most important, if you are striving for insight, interruptions are deadly. A short phone call, email, or text message can redirect your attention and thoughts, and once you are there, it can take a long while to get back. Even the *thought* that some message may be awaiting you can have the same effect. So find a secluded spot, turn off your phone, and don't open the email program on your computer.

These steps are all useful ways to adjust our environment or circumstances to foster the generation of original ideas and insight. In part IV, we'll examine the personal qualities that can aid or deter us and how, contrary to what used to be the traditional wisdom, we can alter our natural style of thinking to better conform to the demands of today's society.

Part IV

Liberating Your Brain

8

How Thought Freezes Over

Building Lives and Candleholders

Jonathan Franzen is living his second life. He'd plotted out his first when he married his Swarthmore College sweetheart. His disheveled graying hair, black plastic glasses, and five o'clock shadow give him a hip, professorial look, which would have been a perfect fit in that original blueprint. "We planned to write literary masterworks that would be published and earn us a reputation," he tells me. "By the time we were in our thirties, we would have good teaching jobs at a nice college and we would live in an old Victorian house and have a couple of kids and just have a good literary life."

Knowing the market for both literature and professors of literature, I found this strangely confident, as if in graduate school I'd made plans to discover a new elementary particle and then settle down to teach at Harvard. But Franzen shows no sign of irony in outlining his thinking. In his young mind, the dream must have seemed attainable. He'd even predetermined the kind of books he would write. "My parents had given me the injunction to do something useful for society," he says. His resulting feeling of responsibility had an enormous influence on the way he viewed his talent. "I had to make the world a better place," he tells me. "And so I imagined that there had to be some sort of social or political critiques embedded in my books—that they had to be about the fate of a city or nation."

As may happen with even the best-laid plans, events didn't work out as he'd envisioned. Franzen's first novel, *The Twenty-Seventh City* (1988), was met with positive reviews, though not spectacular sales. His second book, *Strong Motion* (1992), had little impact and disappointing sales. His career stalled, and, worse, he found it unfulfilling. For Franzen, these were hard times.

Something had to change. To reinvent his career, however, Franzen would have to overcome a phenomenon psychologists call "functional fixedness." The term refers to the difficulty people have envisioning that a tool traditionally used for one purpose can be gainfully enlisted for another. Consider the following classic experiment.

Subjects are given a box of tacks and a book of matches and asked to find a way to attach a candle to a wall so that it burns properly.

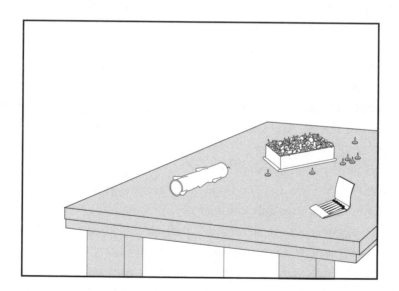

Typically, the subjects attempt to employ the items in the conventional manner. They try tacking the candle to the wall or lighting it to affix it with melted wax. The psychologists had, of course, arranged it so that neither of these obvious approaches would work. The tacks are too short, and the paraffin doesn't bind to the wall. So how can you accomplish the task?

The successful technique is to use the box as a candleholder. You empty it, tack it to the wall, and stand the candle inside it, as shown below. To think of that, you have to look beyond the box's usual role as a receptacle and reimagine it serving an entirely new purpose. Subjects had a difficult time doing that because their familiarity with the intended use of a box interfered with their ability to envision a new use.

That brings us back to Franzen. His tool was his talent, and his fixation arose from that grand vision he'd developed for his life and his art. Before he could restart his career, he had to see that, like the tack box, his gift of writing could be applied in a different manner—to writing a different kind of book.

In the candle experiment, when given a deadline of several minutes, about three-fourths of subjects fail to find the solution. Young children, when given this or similar puzzles by research psychologists, do much better. In one study, so did the members of a hunter-gatherer tribe in the Amazon forest of Ecuador. Both the children and the hunter-gatherers lacked experience in the normal use of the items provided, which enabled them to perform better than those who had preconceptions about their uses based on long familiarity with them.

In life, once on a path, we tend to follow it, for better or worse. What's sad is that if it's the latter, we often accept it anyway—not because we're afraid of change, but because by then we are so accustomed to the way things are that we don't even recognize that they could be different.

In the past few chapters, I've talked about the importance of how we frame a problem or issue, and how we generate new ideas and achieve insight into challenges that are stumping us. In this and the next three chapters, we will look at the other side of that coin—what is holding us back, and how we can overcome it.

The Momentum of Thought

In the psychologists' definition, functional fixedness refers to how our usual modes of thought can constrain the breadth of our new ideas in the context of tool use. But that is just one manifestation of a larger issue in the way the human brain deals with unfamiliar situations. One can call that "the momentum of thought," for, like a mass in Newton's first law of motion, once our minds are set in a direction, they tend to continue in that direction unless acted upon by some outside force. It holds many of us back, keeps us from seeing the changes that would improve our level of satisfaction in life. More generally, it impedes our ability to think of new approaches and imaginative ideas.

In a new circumstance, the momentum of old thinking can condemn you to trying to fit square pegs in round holes. When encountering an unusual challenge, does your mind create an appropriately novel response, or do you stick with familiar ideas and concepts? Do you see a tack box as an object full of potential, or do you see it as merely a receptacle for holding tacks?

A new or changed situation may provide the force needed to alter the direction of your thinking. For some of us, all that is required is a tiny nudge. Others need to run into a brick wall. Franzen ran into the wall. After the failure of *Strong Motion*, his marriage, more than a decade old, began to fail. Then his father became ill with Alzheimer's. The series of events brought years of discouragement and depression. But they also brought him some good, because all the disruptions liberated him from his fixed way of thinking about himself. With regard to his career, he began to see that the tool of his writing talent, like the psychologists' box of tacks, could be employed in a manner he'd never before imagined, and that new freedom completely transformed him as a writer.

"I realized I was trying to be a kind of writer that I wasn't best at being," he says. "So I abandoned my notion about the novel's place in the world and decided I could write about the problems of a set of characters, rather than those of society." When Franzen speaks, his words are measured, and this sounds as much like a statement about the philosophy of literature as it does a personal revelation. But the change he's describing was enormous.

Over time, the operational implications of Franzen's realization fell into place. He stopped worrying about writing for the masses and realized he should write for people who particularly love books. It would be enough if he could give those readers a good time, give them insight into the kinds of issues we all face, and make them feel less alone. "I began to construe my novels as a series of interlocking modules," he says, "each focused on the arc of one character. And I stopped worrying about creating plots with high stakes. My biggest breakthrough was realizing that I could build an entire book around the question 'Will a woman get her family together for Christmas?'"

It was a huge shift in approach, but it worked. In 2001, Franzen published *The Corrections* to great acclaim, and his career has thrived ever since. He became America's bestselling writer of literary fiction, won a National Book Award, and landed on the cover of *Time* with the headline "Great American Novelist."

In the preface to his 1936 book *General Theory of Employment, Interest and Money*, economist John Maynard Keynes wrote, "The ideas which are here expressed so laboriously are extremely simple and should be obvious. The difficulty lies, not in the new ideas, but in escaping from the old ones, which ramify . . . into every corner of our minds." Franzen's success is a parable of liberation, a story of the benefits of elastic thinking, of the potential we can achieve. The lesson is that by letting go of our fixed ways, we can accomplish goals we might never have thought possible.

When Thought Freezes Over

At the turn of the twentieth century, famed physicist Sir James Jeans helped derive a theory of a phenomenon called blackbody radiation. He based his theory on Newton's laws and the well-established theory of electromagnetic forces. It was a beautiful theory, stemming from well-established physics. But when he compared its predictions with the experimental data, the theory failed miserably. Today we know that Jeans's mathematical recipe was sound. It was just not meant to be applied to the ingredients he was cooking with: Newton's laws are not valid for atoms, and it is the motion of atoms that creates blackbody radiation.

About the time Jeans was creating his theory, an obscure physicist named Max Planck cooked up something different, based on an alteration to Newton's laws that he had invented. He called it the *quantum principle*. Unlike Jeans's theory, Planck's novel recipe resulted in predictions that matched the experimental data beautifully. When asked about this, Jeans admitted that Planck's theory worked and his didn't. But, he added, he believed his theory was correct anyway. Had you asked Sir James Jeans a question about almost any topic in physics, you would have gotten a brilliant answer. But with regard to his own failed theory, he sounded like a used car salesman insisting that the transmission isn't really that important.

The political theorist Hannah Arendt defined "frozen thoughts" as deeply held ideas and principles that we long ago developed and no longer question. In Arendt's eyes, the complacent reliance on such accepted "truths" was akin to thoughtlessness, something like the automatic scripted behavior of the mother goose, a computer, or a human operating on autopilot. People operating according to the dictates of frozen thoughts may be powerful processors of information, but they blindly accept ideas that conform to their frozen thoughts and resist accepting nonconforming ideas even when there is ample evidence for them.

Frozen thinking occurs when you have a fixed orientation that determines the way you frame or approach a problem. Our challenge is to turn off that mode of mental operation, to defrost and reexamine our "frozen thoughts" when it is appropriate. Arendt called the kind of thinking we engage in when we rise above frozen thought "critical thinking." To Arendt, who was interested in the origins of evil, thinking critically was a moral imperative. In its absence, a society can go the way of Nazi Germany, a risk that is still present in many countries today. And yet, Arendt noted, a surprising number of people don't think critically. "[The] inability to think [critically] is not stupidity," she wrote. "It can be found in highly intelligent people."

It is ironic that frozen thinking is a particular risk if, like Sir James Jeans, you are an expert at something. When you are an expert, your deep knowledge is obviously of great value in facing the usual challenges of your profession, but your immersion in that body of conventional wisdom can impede you from creating or accepting new ideas, and hamper you when you are confronted with novelty and change.

In my years in science, I have heard many colleagues complain that the experts who referee papers sometimes approach them from a fixed point of view and proceed to misunderstand what they read because they approach the material hurriedly, thinking that they already know what the authors are trying to say. Just as an experienced golfer can have difficulty altering the much-rehearsed stroke that is encoded in his motor cortex, so too may a professional thinker have difficulty shedding the conventional ways of thinking lodged in her prefrontal cortex. Or, as the photographer Dorothea Lange wrote, "To know ahead of time what you are looking for means you're then only photographing your own preconceptions, which is very limiting, and often false."

Frozen thinking has plagued the careers of scientists and ruined the health of many a business, but one context in which it is truly dangerous is medicine, and public health researchers have only recently begun to uncover the ramifications. For example, if you land in the hospital, it's natural to want to be treated by the most experienced physicians on staff. But according to a 2014 study, you'd be better off being treated by the relative novices.

The study appeared in the prestigious *Journal of the American Medical Association (JAMA)*. It examined ten years of data involving tens of thousands of hospital admissions and found that the thirty-day mortality rate among high-risk acute-care patients was a third lower when the top doctors were *out of town*—for example, when they were away at conferences.

The *JAMA* study didn't pinpoint the reasons for the decreased death rate, but the authors explained that most errors made by doctors are connected to a tendency to form opinions quickly, based on prior experience. In cases that are not routine, that can be misleading, because the expert doctors may miss important aspects of the problem that are not consistent with their initial analysis. As a result, although junior doctors may be slower and less confident in treating run-of-the-mill cases, they can be more open-minded in handling unusual cases or treating patients with subtler symptoms.

That alarming finding supported another bold study published in an obscure Israeli medical journal. The question addressed in that research was whether doctors mired in frozen thought might prescribe drugs too readily, and without enough scrutiny of the specific circumstances of individual patients. In particular, a doctor on autopilot might not give due consideration to the interactions of new drugs with the many other drugs a patient is already taking.

To probe that possibility, the scientists enlisted 119 patients in geriatric nursing homes. Their subjects had been taking an average of seven medications each day. With careful monitoring, the researchers discontinued about half the medicines. No patient died or suffered serious side effects from stopping the drugs, and almost all reported an *improvement* in health. Most important, the death rate among those in the study was far lower than that of a control group whose members had continued their medications. It is medical dogma that medications extend life, but that can backfire when doctors are frozen in the textbook approach to all patients.

The kind of elastic thinking that doctors and the rest of us who might be expert in something need to strive for was illustrated well by a simple case study that appeared in another *JAMA* article. A six-year-old boy was brought to the pediatrician for behavioral problems. After speaking to the mother and the boy, the pediatrician concluded that the symptoms indicated a diagnosis of ADHD, and he wrote a referral for psycho-educational testing. Then the mother mentioned in passing that the boy, who was asthmatic, had been coughing a lot lately and using his asthma inhaler more often than usual in order to try to control it. Rather than allow his prior diagnosis to close his mind, the doctor absorbed this new piece of evidence—hyperactivity can be a side effect of the medication in the asthma inhaler. He postponed the testing and prescribed an asthma control medication so that the boy would not have to use the inhaler as frequently. As it turned out, that solved the problem.

Destructive Doctrine

Some of the most tragic but enlightening examples of frozen thinking come from the annals of war. The military is particularly vulnerable to frozen thinking because, in the military, expert and authoritative thought is institutionalized. The military operates according to strict rules dictated from the top according to the generally accepted principles, and passed down through orders to the ranks. "In the military," says General Stanley McChrystal, "we have a doctrine of military operations. The longer you operate under that doctrine, the greater the danger that you become shaped by it."

McChrystal should know. He spent more than thirty years in the army, rising to the rank of four-star general. He finished his career as commander of U.S. and international forces in Afghanistan, and of the Joint Special Operations Command, which put him over the Delta Force, Rangers, and Navy SEALs, the teams that conduct our most secret missions and most of the high-profile operations that make the headlines. Among others, McChrystal oversaw the units that captured Saddam Hussein and tracked down Abu Musab al-Zarqawi, the leader of Al Qaeda in Iraq.

McChrystal is credited with revolutionizing modern warfare with his tactics of invading not just enemy positions, but also their phones and computers, and streamlining the decision-making process required to order such raids—the enemy was not saddled with a heavy bureaucracy, and if we wanted to keep up with them, we couldn't be, either.

As McChrystal's successor, General David Petraeus, told me, today "it is often the side that adapts the fastest that prevails." And so, where other wars were fought on the basis of the lessons learned from conflicts fought decades earlier, successful warfare now requires that you create your theories of battle on the fly. Petraeus, for instance, wrote a "counterinsurgency guidance" document that he kept on his laptop and updated weekly.

One of the challenges for both McChrystal and Petraeus was the need to coax their commanders into embracing this more improvisational approach. McChrystal tells me he understands the hesitancy of those who had trouble with the new ways. He knows there was comfort in obeying the old, accepted military doctrine. You feel you can't go too wrong. After all, doctrine is based on the lessons of history. But to rely on a fixed doctrine is a false comfort, and a dangerous one—it can lead to disaster if conditions change and the doctrine doesn't.

As I talk to McChrystal about historical examples of frozen thinking, we turn to the Yom Kippur War, which began when adjacent Arab states staged a surprise attack on Israel on the Jewish holiday of Yom Kippur, on October 6, 1973. The tale has since become a classic in the fields of political and military psychology, alongside the surprise attack at Pearl Harbor, the German attack on the Soviet Union in June 1941, and the frenzy of miscalculation and misjudgment that led to World War I.

McChrystal talks about the first signs that might have served as a warning to Israel. They came that August, when Israeli military intelligence reported that her northeastern neighbor, Syria, was moving Russian antiaircraft missiles to the border on the Golan Heights. Then, in late September, Syria began a massive mobilization, deploying an unprecedented amount of artillery to the Golan. The movement of one armored brigade in particular should have raised eyebrows. The brigade had been stationed to keep peace at the Syrian city of Homs. Removing it was dangerous, because the city was a hotbed of Islamic opposition to the ruling regime. In fact, a decade later, the Syrian military would be forced to conduct a major operation there, killing an estimated fifteen thousand of the city's inhabitants.

As these events were unfolding to the north, to the south, Egypt was mobilizing reserve soldiers and transporting them to its border with Israel along the Suez Canal. Convoys arrived daily, including several hundred ammunition trucks. Meanwhile, the reserves paved roads in the desert and worked into the night preparing structures that overlooked the Israeli positions, and descents from which they could lower boats that could cross the canal.

By the start of Yom Kippur—a day on the Jewish calendar when

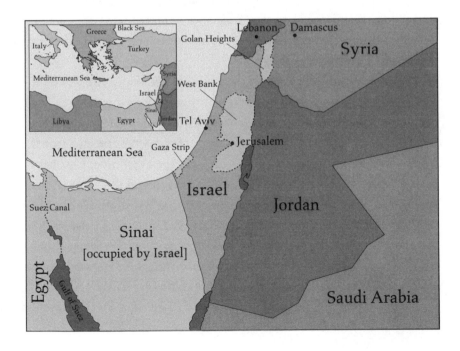

all normal activity ceases and Jews go to their temples and synagogues to pray—Syria had an eight-to-one advantage in tanks, and an even greater advantage in artillery and infantry. Egypt had 100,000 soldiers and 2,000 artillery pieces and heavy mortars on the west bank of the Suez Canal. Israel had 450 men and 44 artillery pieces spread over a hundred miles on her side of it. Why didn't the Israelis see that huge buildup—which they were aware of—as a threat?

"The essence of military deception," McChrystal tells me, "is to figure out what the enemy is preset to believe, and to play to that. Israel didn't connect the dots because they believed that the Arabs did not want to risk losing another war."

The Arabs counted on that Israeli assumption and explained away the buildup as merely a joint training exercise. If it had been, it would have been one of unprecedented scope. But the two Israeli men in military intelligence who were responsible for assessing such threats and communicating them to the leadership—Major General Eli Zeira and Lieutenant Colonel Yona Bandman—didn't grasp what was happening. These highly intelligent, well-trained, and experienced officers accepted the "military exercise" explanation for the events and discounted the likelihood of an attack, despite the fact that the Syrian and Egyptian leaders had on many occasions loudly and publicly declared their goal of destroying Israel.

For each seeming anomaly in the exercise scenario, Zeira and Bandman had an explanation, their frozen thinking blinding them to what should have been obvious. As a result, in the early afternoon of October 6, Israel was faced with a massive surprise attack on two fronts.

The first two days sent Israel reeling. By the evening of October 8, with Arab forces closing in from both the north and the south, Israeli minister of defense Moshe Dayan told Prime Minister Golda Meir that the State of Israel "is going under." But eventually Israel beat the odds. By the time a cease-fire was agreed to on October 24, Israel's forces had progressed to within twenty miles of Damascus and a hundred miles of Cairo, causing the Soviet Union to threaten to send troops to support the Egyptians, and the United States to respond by putting its nuclear forces on a higher state of alert.

After the war, Israel set up a top-level committee—the Agranat Commission—to study how its leaders could have ignored the overwhelming evidence of an impending attack. They concluded that plentiful evidence of an impending attack had been gathered but that the intelligence establishment had misinterpreted it because of their prior beliefs.

The commission concluded that the most important cause of the intelligence failure was the unswerving faith in a doctrine that was so central to their analysis that it was called simply "Ha'Conceptzia" (the Concept). The Concept emerged from intelligence reports detailing secret assessments made by Egypt's leaders after the Six-Day War of 1967, in which the Israeli air force had been decisive in its lightning victory. It asserted that Egypt would not start another war with Israel before gaining air superiority. Since the Israelis had a larger air force than the Arab states, their belief in the Concept translated to strong confidence that the Arabs would not dare attack.

Unfortunately for the Israelis, the Arabs had reinterpreted what air superiority meant. To the Israelis, superiority meant having a bigger air force, but the Arabs believed they had achieved it by acquiring more antiaircraft missiles. "Arab thinking had changed," McChrystal tells me, "and the Israelis hadn't noticed."

Due to the Israelis' adherence to the Concept and their failure to understand that the terms had changed, their intelligence leadership managed the considerable feat of ignoring evidence of an intention to invade that was so clear and obvious, any novice would have spotted it.

McChrystal tells me that the lesson of the Yom Kippur War mimics the challenges we've faced in the Middle East since our invasion to oust Saddam. "We expected the terrorists to have certain limitations. We went in with an almost formulaic approach, like the old Green Bay Packers with their scripted but successful series of plays." But then we ran into Al Qaeda in Iraq, a nimble and decentralized band of terrorists. "They were a constantly changing organism that adapted quickly," McChrystal explains, "and so our strategies soon became ineffective. In that environment, the answers that worked yesterday probably won't work tomorrow. We had to learn to be as flexible as they were. But when you ask a culture that has been very successful to change, it requires some shaking."

Shaking things up is precisely what McChrystal did. "He far exceeded the authority anyone would have presumed he had," one of his colleagues, General James Warner, told me. "He disassembled the organizations and reassembled them. When he was done, he'd collapsed what we would normally consider months-long decision cycles into just hours."

Unfortunately for McChrystal, his penchant for voicing his disagreements with others openly and sometimes crudely got him in trouble with the White House, and he was fired in 2010 by then-president Obama. But he had left his mark. His legacy, as a *Forbes* magazine piece on leadership put it, is that he created "a revolution in warfare that fused intelligence and operations." To McChrystal, though, it was a simple matter of applying elastic thinking. "Weak commanders look for prepackaged answers," he tells me. "Strong leaders adapt."

Handicapping the Expert Brain

If expertise can impede your thinking in situations of novelty or change, then how big is that influence? In the *JAMA* study, researchers suggested that less expert doctors might be superior in diagnosing and treating unusual cases, but the researchers didn't study the relationship between, say, the number of years of experience and the size of that effect. Amazingly, at least in one context, psychologists *were* able to quantify such a relationship, and the magnitude of the influence was startling.

The context the psychologists studied was the game of chess. They began by showing their subjects "board positions" of the type one might find in a chess magazine or book. These are diagrams illustrating where the pieces are, in the midst of a hypothetical game. The board positions are designed so that one player has the upper hand and can force a checkmate if he makes the proper sequence of moves, called a "combination." By "force," I mean that no move his opponent can make will prevent the checkmate. The challenge for the magazine reader is to find the winning combination.

A chess game is not a short affair. If you open a bottle of wine at the start of one, you might have vinegar by the time it's over. Chess aficionados can also discuss at length the elegance of the *way* a game was won. In the experiment, some subjects were shown a board position in which white could win in just one way, through a clever sequence of three moves. I'll call that the "one-solution board." Others were shown a board position in which two different winning combinations were possible—the clever one I just mentioned, plus another that is easier to find, but that chess connoisseurs would consider inelegant. I'll call that the "two-solution board." Would the presence of the easy-to-see, inelegant solution hamper players from discovering the clever, elegant one?

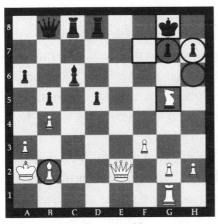 

2-solution problem   1-solution problem

Two board positions: two-solution board (left) and one-solution board (right).
The solutions appear in the endnotes. The crucial squares for the familiar solution
are marked by rectangles (f7, g8, and g5) and the optimal solution by
circles (b2, h6, h7, and g7).

The researchers found that, given enough time, all the players shown the one-solution board would find the clever checkmate. But those presented with the two-solution board had great difficulty finding that same set of moves. Once players had found the familiar, obvious checkmate, they were impeded from seeing the shorter, elegant one, even though they'd been told it existed and searched for a long time.

That is classic frozen thinking, analogous to the other examples I've discussed. What made this study special was that scientists were able to quantify the correlation between how expert the players were and how much "dumber" their expertise made them with regard to discovering the unusual "clever solution."

The key to quantifying the effect is that chess players have a numerical rating system. They have the convenient habit of playing other chess experts of known strength, keeping careful records of who beats whom, and awarding points accordingly. Translated to probability of winning, the system says that if you play someone 200 points above you, your chances of a win are just 25 percent; if you play someone 400 points above you, they are only 9 percent.

When the scientists compared the performances of those at different point levels on the one- and two-position problems, they found that the presence of the extra routine solution produced a dumbing down equivalent to *600 rating points*. That difference is enormous. Translated to intelligence testing, for example, it amounts to a 45-point difference in IQ. That's food for thought: Sometimes when the experts fail, the answer is to call in a novice.

Physicist James Jeans, the Israeli intelligence officers, the top doctors, and the expert chess players were all snared by a similar trap. Whether our knowledge bank encompasses the theories of physics, or strategies of war and peace, or chess maneuvers—or anything else, for that matter—what we *know* can put a constraint on the possibilities we can *imagine*. Having a depth of knowledge is usually desirable, but how can experts battle frozen thinking?

The Benefits of Discord

When psychologists study frozen thinking, they call it "dogmatic cognition." In the psychologist's definition, it is "the tendency to process information in a manner that reinforces the individual's prior opinion or expectation." Zen Buddhism has a concept for a style of thought diametrically opposed to dogmatic cognition. It is called "beginner's mind." It refers to an approach in which you have a lack of preconceptions and perceive even routine situations as if you are encountering them for the first time, without automatically making assumptions based on your past experience. That doesn't mean you discard your expertise, but that you remain open to new experience despite it. Most of us have a cognitive style that falls somewhere between the extremes of a beginner's mind and dogmatic cognition.

The ideal expert in any field is one who has a great breadth and depth of knowledge and yet maintains, to a large extent, a beginner's mind. Unfortunately, gaining expertise can make it more difficult to process new information with an open mind. As one scientist wrote, "Social norms dictate that experts are *entitled* to adopt a relatively dogmatic, closed-minded orientation." As a result, he added, "self-perceptions of high expertise elicit a more closed-minded cognitive style." We probably all know people like that.

Fortunately, psychologists have found that you can move your thinking away from the dogmatic-cognition end of the spectrum. One of the most effective ways is to introduce a little discord to your intellectual interactions.

Consider a study, performed about a half-century ago, by Serge Moscovici, a Holocaust survivor who went on to study group psychology. Moscovici showed two groups of volunteers a sequence of blue slides. In the control group, after each slide he asked the individuals, one by one, to state the color of the slide and estimate its brightness. In the experimental group, he had planted some confederates—actors who called the color "green" rather than blue. Who were they fooling? Nobody—the experimental subjects ignored the deviant responses. When their turns came, they all answered "blue," just as the control group had.

Afterward, Moscovici asked all the volunteers to participate in what he said was a second, unrelated experiment. But it wasn't unrelated. In fact, the earlier experiment was just the setup for the second experiment, which was the one that counted. In that experiment, all the subjects were asked to classify, this time privately and in writing, a series of paint chips as either green or blue, even though their color lay somewhere between those two pure colors.

In this experiment, those who had earlier been in the control group responded quite differently from those who had been in the experimental group. Those who'd been in the experimental group identified as "green" many chips that those who had been in the control group called "blue."

On the color spectrum, the colors from green to blue form a continuum. Starting with green, they proceed through successively more bluish greens, through the greenish blues, and finish at blue. The second experiment was essentially a test of where you draw the line separating greens and blues. Amazingly, the experimental group's exposure to the earlier misidentification of blue as green had shifted their judgment. Compared with the control group, they were more open to seeing a color as green.

Think about that. Not a single subject had been convinced by the earlier misidentification. Yet they *were* influenced by it. They had been subconsciously persuaded to slide the line of demarcation between colors over toward the green. What does this say about human thought? Even if you're not consciously open to considering opposing points of view, with a little exposure, they can affect your thinking.

Other experiments show that dissent can not only sway us with regard to the issue at hand—it can also act to thaw frozen thinking in contexts unrelated to that in which the dissension was voiced. Yes, as distasteful as it is, it is beneficial to talk to people who disagree with us. So if you hate conspiracy theories and run into someone who believes that we faked the moon landing and Einstein plagiarized relativity from his mailman, don't tell him, "Your life is a cruel joke" and walk away. Have tea with him. It can broaden your style of thinking, and it's cheaper than seeing a therapist.

Unfortunately, those who suffer most from dogmatic cognition may be reluctant to listen to opposing opinions. Even worse, if they are in positions of authority, they often punish those who have them. Take the Israeli intelligence officers who overlooked the signs of impending war. Major General Zeira told officers who warned that conflict was likely that they should not expect promotion; Lieutenant Colonel Bandman was famous for rejecting any suggestion that he change even a single word in any document he wrote. It's obvious that allowing dissent—and carefully considering it—could have helped their thinking.

That's one of the benefits reaped by universities and corporations that seek diversity in their students and employees. In addition to whatever brilliant ideas those individuals might bring with them, the mere presence of those with other points of view creates a spirit that encourages liberation from deeply ingrained assumptions and expectations. It promotes the consideration of more options and leads to better decision-making. It builds an atmosphere in which people can better respond to change. That was one of the Agranat Commission's central conclusions: that to avoid the mistakes of frozen thinking, Israeli intelligence needed to restructure in a manner that fostered dissent and invited unconventional ways of viewing situations and events.

As we experience the world, we learn useful facts and valuable lessons, and we form a point of view. Over time, we add to and adjust that point of view, much as we might add on to or update our house over the years. But just as we'd hesitate to add a contemporary wing to an old Victorian home, we resist making changes to the edifice of our worldview if those changes don't seem in harmony with what is already there. And yet, in this rapidly evolving world, that is often what is called for. And so it is one of life's ironic truths that, though we love to be right, we are better off if sometimes people tell us we are wrong.

9

Mental Blocks and Idea Filters

When Believing Means Not Seeing

When I was an impressionable child, my father told me of a Yiddish saying that he'd come to value in his own younger days as an underground fighter during World War II. In English the saying translates to something like "When a worm sits in horseradish, it thinks there's nothing sweeter." To that my father added, "And if the worm sits there long enough, everything looks like horseradish." Those are simple sentiments about the mental barriers that stand in the way of imagining things being different from the way they are, or have always been. For my father, who fought in the resistance during World War II, those barriers were a useful tool when trying to hide fugitives or events from the Nazis.

I pondered that principle of human thinking one night while I was at an academic conference on an old estate in rural England. A bunch of us who'd been drinking got pulled into a late-night Monopoly game. To break the tedium, everyone was talking a great deal, but I decided to use the game to test my father's folk wisdom. In Monopoly, there is a "bank," which is the game box. It holds neat stacks of the various bills of play money, in denominations from $1 to $500. It is common practice to occasionally reach into the bank to make change or to ask someone to reach in to change a bill for you. Everyone was so habituated to seeing players do this that I wondered if they were drowning in horseradish. Would they notice if I practiced a slight variation on the usual change-making?

I decided that when I reached into the bank, I would casually deposit a $20 or a $50, but withdraw a few $100 bills. I did it in plain sight and waited for someone to call me on it. No one did, and so the bank became my ATM. Unfortunately, one can't perform the same feat with a real ATM, so it was one of those experiments scientists sometimes conduct that have no direct practical application.

When I confessed after winning the game, some of my colleagues didn't believe me. They'd become blind to what transpired around them, but they couldn't accept the idea that they had failed to see something so obvious. They'd never have overlooked such larceny, they insisted.

Why didn't they notice? I had observed my fellow players gazing at me as I was making my phony change, so I knew that their eyes had registered my actions, and that their primary visual cortex would have recorded it. But the scene was never passed to their conscious awareness.

Our conscious brains can process about forty to sixty bits per second, roughly the information content of a short sentence. Our unconscious has a much greater capacity. Your visual system, for example, can handle about ten million bits per second. As a result, your primary visual cortex can pass only a small fraction of that to your conscious mind. And so, between your vast unconscious sensory perception and your limited conscious awareness stands a system of "cognitive filters." Those filters make their best guess regarding what is relevant or important. They pass that along to our awareness and censor the rest.

The brains of my fellow Monopoly players didn't flag my actions because one of the factors our filters rely upon in deciding what is important is our expectations. That is rooted in our beliefs, and our past experience of the world. As a result, events that appear routine tend to be rated as less important than novel events or changing circumstances, which might present danger or opportunity. Because my behavior was similar to routine activity, and no deviations were expected, no one noticed.

As the work of Kounios and Beeman suggested, our *ideas* are subject to an analogous filtering process. That's necessary because the human unconscious is extraordinarily good at making associations. Say you ponder whether you want spaghetti for dinner. You associate spaghetti with Bolognese, Bolognese with Bologna, and Bologna with Italy, and soon you are thinking about Botticelli's *Birth of Venus*. You also associate spaghetti with meatballs, and meatballs with sub sandwiches, and that gets you going on nuclear submarines. Such cascades of association lead to showers of new ideas. You could make that Bolognese sauce you once had in Italy. You could fly to Bologna for dinner. You could dine on a nuclear submarine. Some of your ideas are useful, some not, and if your divergent thinking—the production of unusual or original ideas—went unchecked, you would drown in unproductive thoughts.

Our unconscious filters work quickly and effortlessly, with the purpose of suppressing ideas that aren't useful and allowing you to stay focused on the more promising ones. If you are tiling your bathroom, you might consider marble or granite or linoleum, but not coal or Peppermint Patties or newspaper, because your unconscious has eliminated possibilities that appeared to be unhelpful.

The downside of the filtering process is that, just as the Monopoly players' unconscious minds did not choose to bring my actions to anyone's attention, the filtering of ideas sometimes prevents some good ones from getting through—our brains may make unusual and useful associations, only to have them discarded.

The ideal level of filtering would censor Peppermint Pattie tiles while still allowing through unusual possibilities that are worth considering, such as bamboo or rubber. In this chapter we'll examine how our idea filters work, and their role in inhibiting the breakout thinking that is so often required if we are to thrive in society today.

Thinking Outside the Box

Clarence Saunders's fortune came from the grocery business. He got his first job at age nine, working school holidays in a general store. Ten years later, he started selling wholesale food. Then, one day in the late summer of 1916, a department store that wanted to open a grocery asked Saunders to travel from his home in Tennessee to Terre Haute, Indiana, to spy on a store that was rumored to have an innovative design.

Grocery stores in 1916 still operated as they had in the nineteenth century. Despite the fact that manufacturers had by then invented techniques that made possible canned and prepackaged items—so that everything wouldn't have to be stored in vats and bins—stores continued to keep all their goods behind the counter. This meant that customers had to tell the store clerks what they wanted and then wait for them to pick out, price, and bag the goods. During slow hours, store clerks often had little to do. During busy hours, they were overwhelmed, causing long lines and keeping customers waiting. The inefficiencies made the grocery business a bad investment, which was why the department store sent Saunders on a fact-finding mission to see if there was a better way of doing things. But at the store in Terre Haute, Saunders found no magic that would make it more profitable.

On the long, hot train ride home, Saunders stared out at the monotonous fields of wheat and corn, the livestock, the dusty small towns. These were common sights that he had seen many times before and ordinarily would have paid no conscious attention to. Despondent, he was ruminating over the wasted trip when his train slowed alongside a pig farm, where a mother sow stood with her six piglets feeding. It was an unremarkable scene in the horseradish of the midwestern countryside. But to Saunders it was as if someone had shoved in his face a blueprint to save the grocery business. The piglets were serving themselves. Why not let *human* customers help themselves? If you redesigned the grocery stores, customers could pull their own goods from shelves.

Saunders, like everyone else in the grocery business, had thought of it within a fixed point of view that precluded his mind from considering a new system of serving customers. But that scene at the pig farm presented him with the image of a new and better way. Upon his return, he told the department store of the failure of his mission, but he didn't tell them of his vision. Instead, over the next few months, he invented the items he would need to carry it out: shopping baskets, price labels, aisles with shelves and displays, front-of-store checkout counters. All of these items that we take for granted today did not then exist.

Saunders opened his first grocery in 1916, and in 1917 he obtained a patent for his new store design. He named his chain after the piglets, calling his stores "Piggly Wiggly." Within six years, he had twelve hundred Piggly Wigglys, in twenty-nine states, and was a very wealthy man. The chain still exists today, with most of its stores in the South.

Saunders's idea, like the answer to any riddle, is obvious once you've been told it. But if that idea had appeared in the brain of any grocery executive before him, it had apparently not been considered promising enough to be delivered to that individual's conscious awareness. The most brilliant entrepreneurs of the day, anxious to fix the problems in the grocery business—willing, even, to send a spy to steal trade secrets—all failed to invent it themselves.

How does the brain's idea filter work—and how can we overcome the censoring when that is appropriate? Real-life problems like the one Saunders solved are too complex to analyze through controlled scientific experiment. But for studying the mechanics of how people come to think outside the box, scientists have found more abstract problems whose solutions demand essentially the same skill. One of the most studied is a riddle first published in Sam Loyd's *Cyclopedia of Puzzles* a couple of years before Clarence Saunders had his epiphany. Though more than a hundred years old, the problem is still discussed in several new academic papers each month. The challenge, known as the "nine-dot problem," is to connect the dots pictured below with four continuous lines, without retracing or lifting your pencil from the paper.

Despite its simplicity, very few people can solve this problem, even if given hints or a long time to ponder it. In many experiments, the number of successful subjects has been *zero*. It is almost always fewer than one in ten. The problem is so inherently difficult that even among those who have been shown the solution, more than one-third could not reproduce it a week later. Try it yourself. The best most people can come up with are configurations like those below, which all require more than four lines.

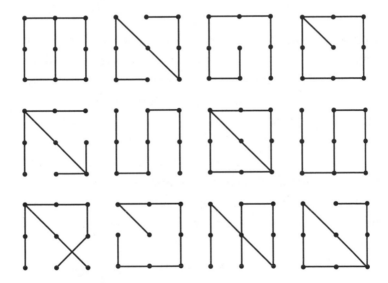

That the solution to the nine-dot problem is not easy is a reflection of the way our brains function. As we've seen, we are not "fresh" observers of the world. What we see (and don't see) is a function of more than what is there. It also depends on what we are accustomed to seeing, and what we expect. If you're accustomed to seeing honest players reaching into the Monopoly bank, you will tend to overlook the one who's cheating. If you are accustomed to having clerks serve customers, it is difficult to arrive at the idea that they could serve themselves. And if you see nine dots arranged in a familiar square array, your brain tends to filter out ideas that involve drawing in the space outside that array. But piercing the imaginary box is what you must do to achieve success in the nine-dot problem. Just look at the solution below:

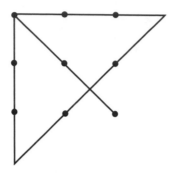

Because the kind of original thinking demanded by the nine-dot problem quite literally requires that you get outside the confines of the box, we call it that—thinking outside the box. One reason psychologists publish so many papers that reference the problem is that by finding ways to improve their subjects' success rate, they shed light on the manner in which our cognitive filters work.

One aid is to supply two strategically placed extra dots, as below. With those additions, although there are more dots to connect, one doesn't have to go beyond the boundary of the diagram to solve the puzzle, and the majority of subjects are able to do it on their first try.

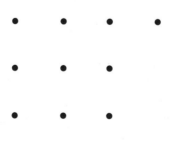

Another way to increase success rates is to draw a spacious box around the dots. Your brain then dispenses with the imaginary box it had defined through the dots and accepts the new, larger box that leaves room for the solution. In other words, the problem can now be solved by thinking "inside the box," which is much easier for all of us:

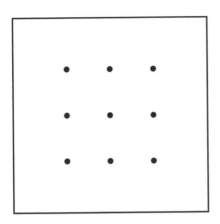 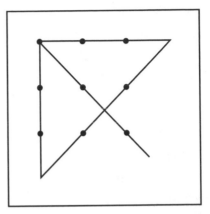

Our brain's cognitive filters are shaped over time. Each day, some neural responses are strengthened, and others suppressed. The result is a brain well adapted to its environment but wired to interpret the world through the lens of what has worked in the past. That allows us to rapidly deal with familiar situations, but it can constrain us from solving others. The latter is what happens in the nine-dot problem: Our sense of geometric borders is so ingrained that our unconscious censors prohibit us from seeing the solution, because it requires violating those borders.

We've seen that altering the way the nine-dot problem is presented can lessen the resistance your unconscious feels to the way the dots must be connected. The key to success in many new challenges, and, more generally, to innovation and imaginative problem-solving, is often to perform mental transformations akin to drawing that box around the dots that made space for the solution.

Recently, scientists have found a physical way to obliterate these unhelpful boundaries to thought—by dialing down the operation of certain structures in their subjects' brains. As we'll see, experiments employing that technique have given scientists a profound view of the physical mechanism through which our mental blocks arise.

Our Idea-Filtering System

In 2012, over a period of eight months, two Australian scientists, "for curiosity," included the nine-dot problem at the end of an unrelated experiment. Of the thirty subjects to whom they administered it, only one was successful. That individual intrigued the researchers, for he had an unusual medical history—he'd suffered a severe traumatic head injury as a child. To the researchers, this was a tantalizing clue. Could his injury have weakened the idea-filtering mechanism in his brain? Unfortunately, it was difficult to pinpoint the exact location of the man's brain damage, so the scientists' study could go no further.

Today, employing a cutting-edge technology, scientists *can* pinpoint and study the structures that form our cognitive filters. It's a technology that allows them to mimic brain damage in healthy people. And that brain damage is—to a scientist—high-quality brain damage: It is well localized; it can be precisely targeted; and, best of all for the subjects, it is transitory.

The method is based on a technique that goes back to ancient Egypt and Greece: applying electric fields to the brain. The ancients did that by placing electric fish (fish that generate electric fields) on the scalp, in an effort to relieve headaches and epilepsy. Whether that worked, no one really knows, but the ancients' batting average was not great. The Egyptians, for example, also used mashed-up mice as a healing salve and crocodile dung as a contraceptive—a method that I imagine would still be effective today, though not for the mystical reasons the Egyptians had in mind.

Today we apply electric fields by employing electromagnetic generators. The generators zap specific neural circuits in the brain with electric or magnetic energy, temporarily disrupting them. Since the zapping is done from outside the skull, the technique is called "transcranial stimulation." It is still being studied as a treatment for certain mental disorders, but it has already been a boon to brain research because of the precision with which it can target specific structures.

For example, in one study, researchers dampened two-inch-square "sponge electrodes" and, using an elastic adhesive bandage to hold them in place, applied them strategically to the scalp of each brave subject, targeting a structure in the brain's cognitive filtering system. Half the subjects then received a small current, which would literally zap that part of their brain into inactivity. The other subjects—the controls—were told they were receiving a zap but didn't. They were all then presented with the nine-dot problem. In an earlier phase of the experiment, the subjects had all attempted to solve the nine-dot problem, and all had failed. The control subjects still had no success. But with the assault on their filters provided by transcranial stimulation, 40 percent of the experimental subjects were now able to solve it.

Through that and other experiments, scientists have begun to piece together the brain's complex system of cognitive filtering. One key structure they've identified is the lateral prefrontal cortex—a mass of brain tissue on the side of the prefrontal cortex region of the frontal lobe. When researchers use transcranial stimulation to interfere with that structure, they find that it improves their subjects' powers of elastic thinking. They become more imaginative and inventive, and more insightful problem-solvers.

Though all mammals have a prefrontal cortex, as I said in chapter 4, only *primates* have a lateral prefrontal cortex—a region defined by its distinctive microscopic structure. It plays a crucial and unique role in human behavior. A key part of your "executive brain," and in particular of its cognitive filtering system, the lateral prefrontal cortex gives us humans our enhanced ability to plan and execute a complex sequence of actions. That function requires an idea filter, because, as I said above, when faced with a situation or goal that demands action, your bottom-up brain gets to work generating possible responses, most of which won't work. It is your lateral prefrontal cortex that follows up by exerting top-down control, biasing you toward some possibilities and withholding others from your conscious consideration. That's why, if you stand at the top of a staircase and want to get to the bottom, you don't think of flapping your wings to fly down or sliding down on your butt (unless you are a child, in which case your prefrontal cortex is not fully developed).

As is true of most brain structures, the function of the lateral prefrontal cortex can be illustrated by the behavior of those in whom it is damaged. My father became such a person after suffering a stroke. Imagine you're on your way to a table at a burger joint and you're very hungry. You walk by a diner who has already been served. Your emotional brain, feeling hunger, might draw you toward snagging something from that diner's plate. But we live in a "civilized" world with rules against that, so your lateral prefrontal cortex, knowing the rules, suppresses that primal idea, and you don't even consider it. One day, however, my father saw some fries on someone's table and grabbed a handful as he walked past, because his stroke-damaged brain didn't censor that possibility or bias him against it.

It is sometimes said that we use only 10 percent of our brain. That's a myth. We use all of it. But we do have untapped brain potential in the sense that there are situations in which it would be advantageous to adjust our filters or alter our mental mode of operation. Transcranial stimulation is one way of doing that. In fact, transcranial-stimulation "thinking caps" have already sprung up for sale to home users. It's not at all clear whether those home devices are effective or, more important, whether they are safe. Many neuroscientists refuse to employ transcranial stimulation, fearing that the devices used by their colleagues in experiments like those I've described could be harming their subjects—and some university ethics committees decline to approve its use. But the day in which such devices will be safe and effective is probably not that far away. We'll explore other means of opening your mind in the next chapter, and they won't require plugging your cortex into an electric power source.

Long Live the Sophomoric

A few years ago my daughter Olivia, then eleven, told my ninety-year-old mother that her "face looks like a raisin." My mother missed the subtle beauty of that metaphor, but she took it with good grace. Certainly, when your brain produces an association between a person's looks and dried fruit, that's best kept to yourself. That's why we admonish our kids to "think before you talk." But if one hears that admonition too often, one might go overboard and get in the habit of preventing ideas from arising in the first place.

To have original thoughts, you have to let the ideas flow first and worry about their quality (or appropriateness) later. And even then, the value of an idea can be difficult to ascertain, for it is one of the ironies of science and the arts that the brilliant and the nutty are not always easily distinguished.

For example, a few years ago, a pair of scientists proposed the strange idea of cold fusion, the ability to create essentially unlimited energy using a simple desktop device. To physicists, that sounded nutty, and it proved to be. But some years before that, other crazy-sounding physicists proposed the strange idea that one could artificially structure composite materials in order to give them bizarre acoustic or optical properties not found in nature—such as being invisible. That also sounded nutty, and they were the subject of snickers and derision. But that idea eventually proved valid, and today their notion of "metamaterials" made from microlattices of metals or plastic is among the hottest topics in science. Scientists have even created small "invisible" objects, though they are invisible only when viewed with a specific color of light (researchers are working on eliminating that restriction).

Or consider Bob Kearns, who created both nutty and brilliant inventions. First, in the 1950s, he created a comb that dispersed its own hair tonic. That sounded nutty. Initially, so did his next invention: the intermittent windshield wiper. Who needs a windshield wiper that stops and goes? It turns out, only people who own cars. Kearns made tens of millions of dollars from that one.

Two-time Nobel laureate Linus Pauling encapsulated the process of innovation when he said, "The way to get good ideas is to get lots of ideas and throw out the bad ones." It's a process full of blind alleys and dead ends. As a result, as Nathan Myhrvold told me, "When people say failure is not an option, it means they are either lying to themselves or doing something boring. When you try to solve an important problem the world has looked at and failed to solve, failure *is* an option, and that is okay." Myhrvold recalled the time an attorney bragged to him that he had never lost a case. "I see," Myhrvold said to the attorney. "So you only do easy stuff!"

As we proceed through life and observe the condemnation of nutty or just incorrect ideas, we may become inhibited. And as we accumulate knowledge and experience, our cognitive filters may strengthen their censorship. But successful scientists, innovators, and artists are usually those who can resist that and maintain the ability to "let go."

One of the most vivid and fascinating accounts of artists "letting go" involves the creation of the classic Indiana Jones movie *Raiders of the Lost Ark*. The film was conceptualized by George Lucas, Steven Spielberg, and screenwriter Lawrence Kasdan in meetings over several days in Los Angeles in 1978. Fortunately, the meetings were recorded, and there still exists a transcript—ninety pages long, single-spaced. When you read it, you are not struck so much by the genius as by the dreadfulness of some of the ideas the iconic filmmakers articulated.

For example, after determining that their hero would need a love interest, the filmmakers decided that when he and the woman meet in the film, they should already have had some history together. They wanted that history to go back about ten years. But they also wanted her to be roughly twenty years old. You'd think that wasn't feasible since it would imply that their "history" happened when she was ten. But with their "lousy idea" filters turned way down, the three movie titans tried to make the math work. Here's how that discussion went:

> LUCAS: We have to get them cemented into a very strong
> relationship. A bond.
> KASDAN: I like it if they already had a relationship at one point.
> Because then you don't have to build it.
> LUCAS: . . . He could have known this little girl when she was just
> a kid. Had an affair with her when she was eleven.
> KASDAN: And he was forty-two.
> LUCAS: He hasn't seen her in twelve years. Now she's twenty-two.
> It's a real strange relationship.
> SPIELBERG: She had better be older than twenty-two.

184

LUCAS: He's thirty-five, and he knew her ten years ago when he was twenty-five and she was only twelve. It would be amusing to make her slightly young at the time.

SPIELBERG: And promiscuous. She came on to him.

LUCAS: Fifteen is right on the edge. I know it's an outrageous idea, but it is interesting . . .

The notion of making Indiana Jones a statutory rapist has to have been one of the most outrageous in the history of rejected ideas, even in Hollywood, where patriarchal conceptions of female sexuality are hardly rare. Fortunately, Spielberg's suggestion that they solve the problem by making Jones's love interest a few years older prevailed.

I talked about the filtering issue with Seth MacFarlane. He was the creator of the popular and long-running television show *Family Guy* and of the *Ted* movies. He has won Emmy awards, been nominated for Grammys (he also sings), and rebooted the classic *Cosmos* science television series, yet, as a *New Yorker* magazine profile put it, he is best known in the popular media as the "No. 1 Offender" in Hollywood, a reputation he received for creating characters that are often racist, sexist, and vulgar. *Rolling Stone* even published a "Hating Seth MacFarlane" timeline. I asked him how his ability to generate new ideas is affected by the threat that what he creates will draw that kind of negative attention.

"It is hard to maintain a mindset in which you are creatively free," he said, "when you feel that if what you say is deemed incorrect, a mob will go after you—that there will be a massive attempt in social media to destroy you. But these days that's what often happens. It doesn't just affect me. It affects everyone in this business, whether they admit it or not. In many ways, social media is the enemy of creativity."

As he and I speak, it is late afternoon, but MacFarlane is at his desk eating lunch, a sparse salad of various sprouts that his personal chef has prepared. "I was eating Triscuits and Devil Dogs for dinner before I hired her," he says. That fare would seem to fit him better. A boyish figure clad in an old T-shirt and baseball cap, he looks more like a wet-behind-the-ears college sophomore than the forty-something man he is, and he seems to still think like a boy, even if his eating habits have evolved. That's what fascinates me about him—but it's what his critics hate. *Entertainment Weekly*, for example, criticized MacFarlane for his "singularly sophomoric mind."

I bring up that criticism because, in creative enterprises, an immature mind is not necessarily a bad trait. Any teacher will tell you that children are not afraid to do or utter crazy things, and in a sense, whenever we relax our inhibitions and allow ourselves to generate an uncensored flood of ideas, we are acting like children.

One reason children are elastic thinkers is that they haven't yet absorbed the full influence of culture or had an abundance of life experience. When you are a child, anything goes. But a few decades later, your dream to someday live in a gingerbread house adorned with pink frosting is not something you'd want to share with your realtor.

Another reason stems from the physical state of a child's brain. As children mature, their more basic cerebral functions develop first— the motor and sensory areas—followed by the regions involved in spatial orientation, speech, and language. Only later come the structures involved in executive function—the frontal lobes. And within the frontal lobes, the prefrontal cortex is the laggard, with the lateral prefrontal cortex maturing last of all. As long as those parts of the brain associated with idea filtering are undeveloped, children remain elastic thinkers, naturally uninhibited. But when we grow up, that spontaneous and unpredictable nature tends to fade. After that, we have to work much harder to think elastically.

Tapping into that childlike imaginative state is what MacFarlane is good at. What he uncovers when he is there is what makes some people dislike him. To others, his unfiltered "sophomoric" humor seduces them into relaxing their adult inhibitions and laughing at political incorrectness that might otherwise make them cringe.

Heroes, from Greek myths to Marvel comics, have special powers. So do each of us, and they change as we move through our lives. The beginner's mind is the Herculean strength of the young, while expertise and the power to know instinctively what works and doesn't work is the spider-sense of the mature. Author and poet Ursula K.

Le Guin is often quoted as having said, "The creative adult is the child who has survived." But the spirit of the child doesn't disappear from our brains; it just becomes more difficult to conjure. The truth is that, within all of us, there exist the neural networks of both a mischievous, imaginative child and a rational, self-censoring adult. The filter in the lateral prefrontal cortex helps decide which of those prevails in any given person. In what follows, we'll examine how the tuning of those filters affects who we are and how we can adjust it.

10

The Good, the Mad, and the Odd

It's a Mad, Mad World

In 1951, the *Proceedings of the Entomological Society of Washington* published a research paper by a gifted University of Massachusetts scientist, Jay Traver. Though best known for her groundbreaking work on mayflies, in this article, Traver discusses in highly technical language and extreme detail how her own body had become stubbornly infected with the common dust mite. She explains that the usual methods employed to get rid of mites, such as shampoos, did not kill the arachnids but merely caused them to migrate to other parts of her body. She describes how, with the aid of several parasite specialists, she applied twenty-two noxious chemicals, ranging from DDT powder to Lysol, in attempts to rid herself of the infestation. None had any effect.

What made this case noteworthy was that house dust mites are not known to colonize humans. They reside in bedding, where they feed on flakes of shed skin. Their presence causes allergic reactions, but they are not parasites. They are also not superbugs—the DDT and other chemicals Jay Traver applied daily should have killed them. Just as mysterious, the mites did not show up, as they should have, in sample scrapings of her skin.

Though her paper was published, scientists eventually concluded that Jay Traver was not suffering from dust mites. One subsequent researcher called her an "authentic mad scientist," as though the other mad scientists he'd run across in the past had all been faking it. Authentic mad scientists, however, are everywhere.

It's not just scientists. One finds a higher-than-average proportion of practitioners displaying odd behavior in all fields that favor elastic thinkers. To name just a few of the more famous "eccentrics": poet and painter William Blake, who was convinced that many of his works were communicated to him through spirits; billionaire entrepreneur Howard Hughes, who had the habit of sitting naked for hours in his "germ-free" room at the Beverly Hills Hotel—on a white leather chair, with a pink napkin draped over his genitals; architect Buckminster Fuller, creator of the geodesic dome, who spent years eating nothing but prunes, Jell-O, steak, and tea and kept a diary in which he made entries every fifteen minutes from 1920 to 1983; singer-songwriter David Bowie, who subsisted on milk and red and green chili peppers throughout his most productive years in the 1970s.

And then there was the brilliant inventor Nikola Tesla. Tesla suffered from hallucinations and unwelcome visions. (He credits his most

Nikola Tesla (1856–1943) at age forty

famous idea, alternating-current electricity, to one of them.) In his later years, he developed an intense fondness for pigeons. On days he was unable to feed them in Bryant Park, in New York City, near where he lived, he would hire a Western Union messenger to do the job. Eventually Tesla became unusually attached to one particular pigeon. It was, in his words, "a beautiful bird, pure white with gray tips on its wings . . . a female." He told a science writer for *The New York Times* that the pigeon "understood me and I understood her. I loved that pigeon. Yes, I loved her as a man loves a woman, and she loved me . . . That pigeon was the joy of my life. If she needed me, nothing else mattered." It was the kind of love we all seek, except for the beak and feathers.

Such stories, about both the famous and the not so famous, are common. Are they just amusing anecdotes, or is there a meaningful link between a tendency toward eccentric behavior and a capacity for elastic thinking?

The first progress toward answers to such questions came in the 1960s, in work by behavioral geneticist Leonard Heston. Curious about the inherited component of schizophrenia, Heston studied children who had been given up for adoption by schizophrenic mothers. To his surprise, he found that half of the *healthy* children of these mothers were both unusually artistic and unusually eccentric. They were not schizophrenic, but they "possessed artistic talents and demonstrated imaginative adaptations to life which were uncommon in the control group," he wrote. That suggested that watered-down schizophrenia might be beneficial—that is, that there was some small inherited dose of schizophrenia that endowed these children with a tendency toward both elastic thinking and nonconformist behavior. If so, what does a "dose" of schizophrenia mean, and how do you measure it?

Measuring Doses of Madness

Psychologists coined the term "schizotypy" to describe a constellation of personality traits such as those the children of schizophrenics seem to have inherited. People with a schizotypic personality may fall anywhere on the spectrum between a mild "dose" of schizophrenic-type traits and full-blown schizophrenia. Over the years, psychologists have developed various personality questionnaires to measure the size of the dose by assessing where people fall on the spectrum. Below is an example of one of these questionnaires. If you want to test yourself, simply answer the following twenty-two statements/questions with a yes or a no, and then tally the number of "yes" responses.

1. ___ People sometimes find me aloof and distant.
2. ___ Have you ever had the sense that some person or force is around you, even though you cannot see anyone?
3. ___ People sometimes comment on my unusual mannerisms and habits.
4. ___ Are you sometimes sure that other people can tell what you are thinking?
5. ___ Have you ever noticed a common event or object that seemed to be a special sign for you?
6. ___ Some people think that I am a very bizarre person.
7. ___ I feel I have to be on my guard even with friends.
8. ___ Some people find me a bit vague and elusive during a conversation.
9. ___ Do you often pick up hidden threats or put-downs from what people say or do?
10. ___ When shopping, do you get the feeling that other people are taking notice of you?
11. ___ I feel very uncomfortable in social situations involving unfamiliar people.
12. ___ Have you had experiences with astrology, seeing the future, UFOs, ESP, or a sixth sense?
13. ___ I sometimes use words in unusual ways.
14. ___ Have you found that it is best not to let other people know too much about you?
15. ___ I tend to keep in the background on social occasions.
16. ___ Do you ever suddenly feel distracted by distant sounds that you are not normally aware of?

17. ___ Do you often have to keep an eye out to stop people from taking advantage of you?
18. ___ Do you feel that you are unable to get "close" to people?
19. ___ I am an odd, unusual person.
20. ___ I find it hard to communicate clearly what I want to say to people.
21. ___ I feel very uneasy talking to people I do not know well.
22. ___ I tend to keep my feelings to myself.

Number of "yes" responses: ___

In a study of about 1,700 subjects to whom this test was administered, the average score—the number of yes responses—was about six. If you responded yes to two or fewer of these statements and questions, you are in roughly the bottom quarter of the population. If you answered yes to thirteen or more, you are high on the scale, in roughly the top 10 percent. The questionnaires suggest the scientists were on the right track. Over the years, those who scored high on such tests tended to be both eccentric and gifted at elastic thinking skills, especially divergent thinking.

Once research in the 1960s and 1970s connected the schizotypic personality type to elastic thinking and eccentricity, psychologists focused on determining the areas of the brain that are responsible for those qualities. It took decades for imaging technology to develop to the point that it could shed light on that mystery. Even then, the issue proved challenging, because although the ideas and behavior of people who register high in schizotypy may appear to be distinctly peculiar, when you look at brain activity, the characteristics of high schizotypy can be subtle. Recently, however, researchers have been able to fine-tune their studies, and the verdict they arrived at might not surprise you: The eccentric/elastic connection arises from decreased activity in the brain's cognitive filtering system, which I discussed in the last chapter.

A lax cognitive filter promotes a high level of schizotypy, a tendency toward original thinking and nonconformist behavior, while a stringent filter produces what psychologists call *cognitive inhibition*, leading to conventional thought and action. If you scored high on the schizotypy scale, you might have an easier-than-average time in this frantic era. That's because those who score high are especially well adapted to new or changing situations. At the highest end of the scale, however, people may have difficulty being coherent.

Just look at mathematician John Nash, about whom the book *A Beautiful Mind* was written. Nash had a schizotypic personality, and cognitive filters tuned low enough that he generated a variety of highly imaginative ideas, such as those on game theory that led to his winning a Nobel Prize. Unfortunately, after doing his groundbreaking research, Nash fell into a long period of full-fledged schizophrenia, during which he was unable not just to work, but to function normally. Any brilliant mathematical ideas he might have had in that period were lost in a shower of wild ones.

That the bizarre and the brilliant often have the same source is illustrated by an exchange Nash had after he eventually recovered. During the period of his illness, he had believed that aliens from outer space had recruited him to save our world. When he was well again, a curious mathematician friend asked him how he could have believed that "crazy" idea. "Because the ideas I had about supernatural beings came to me the same way that my mathematical ideas did," Nash said. "So I took them seriously."

Nash was an extreme case, but imaging studies show that people who believe in other odd ideas, such as telepathy, magical rituals, and amulets of luck, have unusually little activity in their lateral prefrontal cortex and other filtering circuitry. We can even correlate the ebb and flow of that tendency to changes in the brain over one's lifespan: Belief in the supernatural declines as children mature and their lateral prefrontal cortex becomes more fully developed; conversely, in old age, as the vigor of the lateral prefrontal cortex declines and cognitive inhibition decreases, belief in the supernatural increases.

Many of our greatest thinkers seem to have had minds on the high end of the schizotypy scale. Those who have consistently produced original ideas have often also been original, sometimes even bizarre, in their conduct, their grooming and dress, and their relationships. They might even be people who fall in love with pigeons or talk to aliens. In such people, the degree of cognitive inhibition is high enough for them to function, but low enough to allow them to entertain ideas most others would deem inappropriate—including, at times, ideas that change the world.

Elastic Personalities, from the Arts to Science

Different creative pursuits require varying degrees of unconscious elastic thinking, in combination with varying degrees of the conscious ability to modulate it and shape it through analytical thinking. In music, for example, at one end of the creative spectrum are improvisational artists, such as jazz musicians. They have to be peculiarly talented at lowering their inhibitions and letting in their unconsciously generated ideas. And although the process of learning the fundamentals of jazz would require a high degree of analytical thought, that thinking style is not as big a factor during the performance. On the other end of the spectrum are those who compose complex forms, such as a symphony or concerto, that require not just imagination but also careful planning and exacting editing. We know, for example, through his letters and the reports of others, that even Mozart's creations did not appear spontaneously, wholly formed in his consciousness, as the myths about him portray. Instead, he spent long, arduous hours analyzing and reworking the ideas that arose in his unconscious, much as a scientist does when producing a theory from a germ of insight. In Mozart's own words: "I immerse myself in music . . . I think about it all day long—I like experimenting—studying—reflecting . . ."

There is not a perfect parallel between the types of thinking required for success in different creative fields and the personalities of those who practice them, but as the anecdotes I quoted at the start of the chapter suggest, there is some degree of verifiable correlation. In one study, Geoffrey Wills, a psychologist and former professional musician from Greater Manchester, England, investigated the biographies of forty world-renowned pioneers from the "golden era" of improvisational jazz (1945–1960).

Wills found that not only were the jazz pioneers nonconforming, but, on a personal level, they were reckless far beyond what one observes even in other creative fields. For example, Chet Baker was a drug addict whose favorite drug experience was "the kind of high that scares other people to death," the same speedball high, from a mixture of cocaine and heroin, that Timothy Treadwell—and John Belushi—were fond of. Charlie Parker consumed enormous quantities of food and was known to drink sixteen double whiskies in a two-hour period. Miles Davis misused a variety of substances, had many sexual relationships, and had a liking for orgies and voyeurism. Several other greats had a love of fast sports cars, and Scott LaFaro, a notoriously reckless driver, died in a car accident at the age of twenty-five. The excessive sensation-seeking is so common that reading Wills's treatise gets tedious as he details the lives of those I've mentioned, as well as Art Pepper, Stan Getz, Serge Chaloff, and Dexter Gordon, to name a few more.

If jazz pioneers were an especially imprudent group, among professions that reward elastic thinking, science is a field at the opposite end of the spectrum. In science, the ideas you generate have to be more than beautiful or unusual. They have to agree with the results of experiments.

A musician might play to sellout crowds at some basement venue in lower Manhattan even though his work sounds to many people like a raccoon scratching at a blackboard. But a scientist's recipe for turning mercury into gold either works or it doesn't.* As a result, elastic thinking is important in science, but at least as important is an additional skill: an equally strong ability to tame the unrestrained generation of new ideas and to challenge and develop them through analytical thinking.

In science, it's difficult to be successful if you have an "anything

---

* In 1941, scientists really did convert mercury to gold—the alchemists' dream—by bombarding that metal with neutrons in a nuclear reactor.

goes" personality like those jazz greats. And so, people who are successful in science might be eccentric or "mad," but they are usually outliers in less extreme and dangerous ways. Among the scientists I've known personally were an experimental physicist who lunched at the university cafeteria every day but ate only the condiments, a middle-aged neuroscience professor with orange hair and an Apple tattoo, a physics professor obsessed with snowflakes, and a Nobel Prize winner obsessed with banjos. And then there are the more famous examples, such as Albert Einstein, who'd pick cigarette butts up off the street to sniff them after his doctor forbade him from smoking his pipe, and Isaac Newton, who conducted a mathematical analysis of the Bible, looking for coded hints about the end of the world. These great scientists were elastic thinkers, but, in both their professional and personal lives, they employed their executive brains to moderate their behavior more than the pioneering musicians I described.

Though different professions encourage different thinking styles, whether musician, scientist, or original thinker in some other realm, there is a need for a modicum of ordered, analytical thought to transform novel ideas into a creative product—one that is useful, attractive, harmonious, or otherwise compelling. Psychologists believe that one of the key differences between people with schizotypic personalities and those who suffer from actual schizophrenia lies in their ability to focus and, more generally, to apply that kind of orderly, analytical intelligence. Those with a higher IQ seem better able to hold in mind the barrage of odd thoughts that typically emerge from a lowering of cognitive inhibition without becoming dysfunctional in human society. The difficulty of shaping and developing ideas is why, Nash aside, schizophrenics and others with severe psychiatric disorders are *not* well represented in either the arts or the sciences.

The Dr. Jekyll and Mr. Hyde Inside

Growing up in the 1940s, Judith Sussman always sought outlets for her imagination. Sometimes that meant playing with dolls; sometimes it was dancing; sometimes it was simply walking around for hours holding a balloon and making up stories and characters. In the 1950s, the girl with the balloon became a student at NYU, where she met a man with a different mindset and an affinity for analytical thought—a future lawyer. By the 1960s she had settled down and become a housewife with two kids. She soon had a house with many rooms, but no space for the ideas that were always simmering in her elastic mind. While part of her was blooming—she loved being a mother—that other part of her had withered. "I became miserable with my role," she told me. "I didn't have a burning desire to do anything in particular; I just knew I was desperate to be creative again. I couldn't let go of that part of me." That was when she decided to start writing.

Free time was scarce, but Sussman made writing a priority, one ranked almost as high as doing the laundry or preparing the tuna casserole for dinner. She noticed that, to her husband, her new focus felt subversive. He had married a sensible woman who was now betraying him. Her friends weren't supportive, either—this wasn't an era with much tolerance for disaffected homemakers. Nor did she receive encouragement from the editors to whom she sent her material. "I cried when I got my first rejection letter," she said. "And I kept getting rejections for two years."

But Sussman kept writing, and she published her first book in 1969, under her married name, Judy Blume. (She and lawyer John Blume would divorce in 1976.) In the decades that followed, her young adult fiction and four adult novels became huge best sellers, including several that reached number one on the *New York Times* list. Her books have sold more than ten million copies while attracting dozens of literary awards, bringing her a rare combination of commercial and critical success.

Why did Blume stick with it despite the difficulties, the lack of support, the toll it took on her marriage? "Once I began to write," she told me, "I suddenly was anxious to get up in the morning. Writing saved me in those years. Because imagination is something I need in my life. I need it to be healthy. I need it to live. It's part of me."

William James and Sigmund Freud would have understood Judy Blume. Though they knew nothing of the top-down-versus-bottom-up competition within our heads, James and Freud argued that both rigidly analytical and imaginative elastic modes of thought are essential parts of all of us. We are all, in a sense, two thinkers in one.

Consider the following experiment. Researchers asked subjects to analyze the truth of various syllogisms while the researchers imaged their brains in an fMRI machine. Some of the syllogisms were abstract, of the type "All A are B. All B are C. Therefore, all A are C." Others carried meaning, such as "All dogs are pets. All pets are furry. Therefore, all dogs are furry."

From the point of view of pure logic, these syllogisms are identical. The distinction that, in the latter syllogism, the letter "A" has been replaced by a string of letters ("dogs") is of no importance. To our associative brain, however, there is a world of difference. The letter "A" is just the letter "A," but to the word "dogs" is attached a whole catalog of meanings and feelings, all dependent upon who we are as individual people.

A computer would assess the truth of both of the above syllogisms using the same analytic thinking, since that is the kind of thinking it is able to do. And you might think that humans would, too, since these syllogisms have an identical logical structure. But, in fact, the human brain approaches the two syllogisms quite differently. When the experimental subjects judged the truthfulness of syllogisms involving only abstract letters, they used one network of neural structures, and when they judged the syllogisms of meaningful words, they employed another. The precise makeup of those networks is not important to us here. What *is* important is that they are *different*.

Within each of us are two distinct thinkers, both a logician and a poet, competitors out of whose struggle emerge our thoughts and ideas. We can all switch between the mode of thought in which we spontaneously generate original ideas and that in which we rationally scrutinize them, and our success hinges in part upon our ability to shift modes as needed.

In talking to Blume, I got the feeling that if there is one aspect of her existence that she is keenly aware of, it is that capacity for switching between those two distinct modes of thought. Her usual thinking is neat and orderly. But when she writes her novels, Blume says, "it is like I'm another person. I write because there is that other somebody inside me. And it has to express itself. But when I read one of my books after it's been published, I often think, *Did I really write that?*" I know what she means.

11

Liberation

Let's Go Get Stoned

Some years ago, a scientist wrote an essay about his early experiences with marijuana. In his twenties when he had the experiences he was describing, he'd experimented with pot a few times before but had felt nothing. Now he was lying on his back in a friend's living room, trying again. His eyes idly explored the shadows that a plant cast onto the ceiling. Suddenly it struck him: The shadows took the form of a car. Not a generic car, but a miniature Volkswagen, exhibiting intricate detail. He could even make out the hubcaps and the trunk latch. Could there really be a car on the ceiling? That crazy idea survived the cognitive filtering that would have censored it had he been sober. But though it popped into his consciousness, his analytical brain told him it was an illusion. He must finally be high, he reasoned.

The young scientist said that was the moment he discovered that he liked getting high. To me, it wasn't much of a revelation—something akin to when I figured out I preferred chocolate milkshakes to the broiled liver my mother used to make. But in his day, the received wisdom on marijuana was almost universally negative. Smoking marijuana was also, of course, illegal. And while university tenure committees favor those researchers who add to our knowledge of the universe, the discovery that smoking pot is fun isn't what they have in mind.

So when this scientist eventually penned his essay about cannabis, he did it anonymously, as Mr. X, to protect his budding academic career.

Being a scientist, he took note in his essay of what a marijuana high meant. As he described it, he was having perceptions and making associations that in everyday life would seem bizarre, but that in this state felt perfectly reasonable—like John Nash's aliens. Marijuana had enhanced his ability at elastic thinking, and through his cannabis experience, he wrote, he'd come to understand the minds of thinkers we call mad.

He also felt he appreciated music and art in a way he never had before, and had "a feeling of communion with my surroundings, both animate and inanimate." There was even a "religious aspect" to his highs—and a sensual aspect. "Free-associating . . . has produced a very rich array of insights . . . Cannabis also enhances the enjoyment of sex—on the one hand it gives an exquisite sensitivity, but on the other hand it postpones orgasm: in part by distracting me with the profusion of images passing before my eyes."

A touch of madness, the making of unusual associations, the feeling of being in touch with a world beyond the everyday, the heightening of artistic sensibility, a susceptibility to distraction—this scientist's careful description of his drug-induced highs, written in 1969, has striking similarities to the schizotypal personality scientists are just now beginning to understand.

For most of history, we didn't have the technology to decipher how mind-altering substances affect elastic thinking, and even when we did, their illegal status discouraged researchers from studying them. One of the few early studies, published in the prestigious journal *Nature* in 1970, was carried out by a psychologist at the University of California, Davis, a top agricultural school not far from what were then probably the country's largest center of both marijuana use and cultivation. In that study, the researcher distributed questionnaires to 153 potheads, or, as he called them, "experienced marijuana users." The questionnaires asked them to describe the experience and then tabulated the most common responses.

Reading them today, one notes, as did the scientist in his essay, a marked correspondence between the effects of marijuana and an enhanced ability at elastic thinking—at skills such as idea generation, divergent thinking, and integrative thinking, which are aided by the opening of our cognitive filters. For example, a few of the most repeated sentiments were:

"The ideas that come to my mind are much more original."

"I think about things in ways which are intuitively correct, but which do not follow the rules of logic."

"Commonplace sayings or conversations seem to have new meanings."

"Spontaneously, insights about myself . . . come to mind."

"I am more willing to accept contradictions between two ideas."

A friend of mine recently said he wanted to live a healthier life, so he planned to drink less and smoke more pot. His remark reflected the trend toward marijuana acceptance and decriminalization that is now sweeping the Western world. That new social attitude has finally resulted in an uptick in the number of experiments that investigate the anecdotal reports on the benefits of the drug.

In one of those, a 2012 study, 160 cannabis users were recruited and asked to attend two experimental sessions. For one session, they were asked to refrain from smoking for at least twenty-four hours prior to attendance, a condition verified by taking a saliva sample. For the other session, they were asked to bring their own marijuana and smoke it in the lab.

On both days, the subjects were given a battery of tests to measure elastic thinking. For example, fluency was tested by asking subjects to name as many four-legged animals or fruits as they could think of in sixty seconds, and divergent thinking was probed by asking them to generate a word related to all three words in a word triad—the same CRA challenge that Kounios and Beeman employed.

The results were fascinating: Those who did well on the tests when they were sober were unaffected by the marijuana. But those who'd done poorly when sober improved under its influence. In fact, the subjects who, when sober, tested low in divergent thinking, when high, did just as well as the others. The marijuana had boosted their originality of thought. In those skills, as the scientists put it, "smoking cannabis in a naturalistic setting induced significant increases" in schizotypic traits.

That marijuana caused that response is not surprising, once you know how it affects the brain. The active ingredient in marijuana, a chemical called THC, is known to suppress the function of the brain's prefrontal lobe filters. Apparently, those who did well on the elastic thinking test when sober had their filters set naturally low, so there wasn't a great deal of improvement to be had by adjusting them. But the others had more room to improve, and the THC accomplished that. In that sense, marijuana is an elastic thinking equalizer—it allows you to reach toward your maximum potential, though it does little for you if you are already there.

The anonymous scientist finished his essay with the words "The illegality of cannabis is outrageous, an impediment to full utilization of a drug which helps produce the serenity and insight, sensitivity and fellowship so desperately needed in this increasingly mad and dangerous world." He died a couple of decades before his wish would begin to come to pass. His name was Carl Sagan.

In Wine There Is Truth; Also in Vodka

Sagan was right about the benefits of marijuana. But like all drugs, it can have negative side effects. Particularly worrisome is the fact that, if you begin at an elevated level of schizotypy, the use of marijuana can push you over the threshold into psychosis. That may be what happened to Brian Wilson, the leader and co-founder of the Beach Boys. Wilson was one of the most innovative and influential musicians of the twentieth century. His unorthodox approach incorporated the textures of orchestral music into pop compositions, leading to more than two dozen Top 40 hits in the 1960s. His work inspired his contemporaries and energized the California music scene to such an extent that it supplanted New York as the center of popular music. Even his production technique was revolutionary—he used the recording sessions themselves to experiment and create unique arrangements and instrumentation. Today that is called "playing the studio," and it is commonplace, but in the early 1960s it was unheard of.

Wilson began using marijuana recreationally in 1964. Soon afterward, he began to use it for creative purposes. He credited the drug's influence with inspiring him to abandon the simpler conventional rock arrangements and develop his trademark style. But in 1963, Wilson had begun hearing indistinct voices, and after he began using marijuana, his symptoms worsened considerably. He became obsessed with tiny details. Not important details, such as what brand of lemon oil to rub onto the fret board of his bass guitar or whether his accountant was complying with all the tax laws, but pointless details, such as the number of tiles on the floor or the number of peas on his plate. By 1966 he would conduct interviews only in his home swimming pool, convinced that his house was crawling with hidden recording devices.

In 1982, Wilson was diagnosed with schizoaffective disorder. That's an illness in which the victim suffers from elements of both schizophrenia and bipolar disorder, and one that may have been triggered by his heavy marijuana use. We'll never know how Wilson's illness would have progressed if he'd not used marijuana, but his is a cautionary tale. Although marijuana can be helpful for manipulating the balance of forces in your brain, in certain people it can possibly be dangerous.

That's true, too, of another chemical that many eminent artists, musicians, and writers have claimed played a role in their success: alcohol. As musician Frank Varano said, "On some days, my head is filled with such wild and original thoughts that I can barely utter a word. On other days, the liquor store is closed." Such testimonials go back at least as far as 424 B.C., when Aristophanes wrote, in his play *The Knights*, "When men drink, then they are rich and successful . . . Quickly, bring me a beaker of wine, so that I may wet my mind and say something clever."

Recent science seems to confirm that alcohol can have beneficial effects on elastic thinking. For example, in a 2012 study that paralleled the marijuana study done that same year, forty social drinkers in their twenties were recruited through Craigslist. Half were served enough vodka and cranberry juice to bring them to the border of being legally drunk. The others drank just cranberry juice. They were all then given problems whose solutions required elastic thinking. The drunk subjects solved about 60 percent of the problems; the sober ones, 40 percent. What's more, the tipsy students completed the test faster.

The problem with alcohol as a thinking aid is that while the defocusing it provides can loosen the thought processes, they can easily become so loose that they fall off their tracks. The same is true of marijuana. In both cases, the trade-off is much like that in schizotypy versus schizophrenia. Having a drink or two, or a hit of pot, while formulating your business strategy could increase the breadth of ideas that come to you—but if you are too far gone, those ideas might prove useless or incoherent.

Another popular area of drug research today is psychedelics. A few scientists studied the effects of LSD in the 1960s, but although psychedelics are among the least harmful and least addictive "recreational" drugs, almost all such substances were criminalized virtually worldwide in 1971 by the United Nations Convention on Psychotropic Substances. As a result, though the treaty allowed exceptions for scientific or medical purposes, for decades there was effectively no further research. In recent years, however, with the loosening of social attitudes regarding drugs, scientific inquiry into psychedelics has also resumed, and with renewed vigor.

The emerging picture is fascinating—scientists are beginning to connect the anecdotal reports of the psychedelic experience to specific structures and processes in the brain. For example, LSD and psilocybin ("magic" mushroom) users often feel a profound "self-transcendence," a diminished sense of ego, as if the border between themselves and the external world "is dissolving." A group at Oxford made the anatomical connection by administering those psychedelics intravenously and then imaging their subjects' brains by way of fMRI.

The Oxford researchers found that LSD and psilocybin affect elements of the default network. That's the system of structures we encountered in chapter 6 that becomes active when the executive brain is not directing our thought processes. The default network plays a key role in our mind's internal conversations, which help develop and reinforce our sense of self, so the drugs' connection to a diminished awareness of ego is not unexpected. But we also saw in chapter 6 that the default mode plays an important role in elastic thinking. And so the Oxford research raises the question of whether LSD and psilocybin improve or inhibit elastic thinking. Work on the issue is still in progress.

One psychedelic whose effect on elastic thinking is better understood is ayahuasca, a South American psychotropic plant tea made from jungle vines by natives along the Amazon. Several writers, including Chilean American novelist Isabel Allende, have spoken about the effects of an ayahuasca high on their work. Allende, whose books have sold more than fifty million copies and been translated into nearly thirty languages, steeped herself in the foul-tasting brew as a way out of writer's block. For her, it was a transformative experience that loosened up her mind and started the ideas flowing again. "It was the most intense, out-of-my-mind experience that I have ever had," she said. "It was very revealing and very important and opened up a lot of spaces inside me."

Subjects start to feel ayahuasca's effects forty-five to sixty minutes after forcing down the tea. They report seeing visions, perceiving intense emotions, and experiencing a noticeable increase in mental fluency—they generate ideas at an increased pace, especially when they close their eyes. More important, the ideas that come to them are more varied than usual—these subjects perform spectacularly well on tests of divergent thinking. But if aspects of their elastic thought processes are enhanced, ayahuasca, like other drugs, is a double-edged sword: The improvement in elastic thinking comes at the expense of their capacity for analytical thought.

How do a few sips of awful tea exert such a wide-ranging effect on how we think? In chapter 4 I talked about the neural hierarchies in our cortex. At the highest level in each hemisphere are the lobes, which are made of various modules, which in turn are made of submodules, in a scheme that can be followed all the way down to individual neurons. The 180 modules and submodules we've identified thus far send and receive signals via a complex web of neural wiring. The magical offshoot of all that is a flow of information that combines flexible bottom-up and executive top-down processing. Ayahuasca seems to function by interfering with those information flows, reducing top-down control and enhancing the influence of bottom-up processes.

One consequence of that is a loosening of the cognitive grip exerted by the prefrontal cortex. Compared with the effects of marijuana and alcohol, the disruption of the usual paths of neural signal traffic caused by ayahuasca has a far broader and deeper effect, profoundly modifying the user's perception, experience of reality, and even, as with LSD and psilocybin, the sense of self.

More research is needed to elucidate in greater detail the mechanism through which ayahuasca acts. With further research, though, pills to enhance elastic thinking might not be very far off. Some people, especially in Silicon Valley, are already using home-designed "performance psychedelics," such as microdoses of LSD. Such drugs would be the natural partner to the analytical-thinking focus-boosters like Vyvanse and Adderall, which, though they may be addictive, are so common on college campuses, and the memory-boosting pills being developed to help Alzheimer's patients.

Perhaps sometime in the future we will have available a safe and balanced cocktail of such drugs to enhance our overall intelligence. If that were possible, those drugs would certainly be controversial. Some would oppose their use because they are against all mind-altering drugs. Others would point out that they give an unfair advantage to those who can afford them, or that they might have harmful side effects. On the other hand, to raise human intelligence could lead to great scientific and medical discoveries, and innovations that could make life better for everyone.

Whatever future research holds, don't go looking for ayahuasca pills just yet, for the disruption of mental hierarchies produced by the drug exerts a powerful and impractical downside. Allende said she faced demons and saw herself as a terrified four-year-old, curled up on the floor, shivering, retching, and muttering for two days. "I think I went through an experience of death at a certain point," she said. "I was no longer a body or a soul or a spirit or anything. There was just a total, absolute void that you cannot even describe." The ayahuasca slayed her writer's block. But, she concluded, "I don't ever want to do it again."

The Silver Lining of Fatigue

We've seen that drugs and alcohol can enhance elastic thinking by weakening our cognitive filters. Luckily, there are also more natural ways to liberate your elastic mind. In 2015, a group of researchers in France showed, for example, that the simple act of exhausting your executive brain before you start pondering a challenging intellectual issue can unleash your elastic brain to mount a more effective attack.

The French scientists fatigued their subjects' executive brains by putting them through a tedious exercise called the Simon task. In it, participants are shown a set of left- and right-pointing arrows on a computer screen, one of which is always positioned at the screen's center. Subjects are instructed to press the left or right arrow key on the keyboard, according to the direction in which that central arrow points.

The key to the experiment is that, in order to focus on the central arrow, the subjects must suppress the influence of the other arrows. That suppression is accomplished by the subjects' prefrontal cortex, and to perform that task over and over without a break for forty minutes, as the subjects were asked to do, is mentally exhausting.

After the Simon task had dulled the subjects' executive faculties, the researchers presented them with a test of their elastic thinking. The subjects were given a few minutes to imagine as many uses as they could for a set of household objects, like a bucket, a newspaper, and a brick. Their answers were scored according to criteria such as the total number of uses the subject was able to imagine and the originality of each idea (as judged by the number of other subjects who had also thought of that use). The scores were then compared with those of a control group who had not first engaged in the Simon task.

The researchers found that when a subject's capacity for executive function was depleted, both the total number of imagined uses and their originality were significantly greater. The lesson is that, though we expect our best thinking time to be when we are fresh, our *elastic* thinking capacity may be highest when we feel "burnt out." That's good to know when scheduling your tasks—you could be better at generating imaginative ideas if you do that kind of thinking after working on a chore that involves a period of tedious, focused effort that strains your powers of concentration.

The French study also raises a question about our personal rhythms. Not everyone regularly feels their sharpest at a particular time of day, but for many, the labels "morning person" and "evening person" are well deserved—studies confirm that our bodily processes, like heart rate, temperature, alertness, and the executive functioning of our prefrontal cortex, indeed follow regular daily rhythms. These vary from person to person, governed by a cluster of about twenty thousand neurons in our hypothalamus, just above our brain stem. So if you find that you can sit down, focus, and grind through your spreadsheets, professional reading, and other analytical work with maximum efficiency in either the morning or the evening, there is a good physiological explanation. But the French study suggests a new wrinkle to that: Your elastic thinking ability may peak at the other end of the day, when your analytical powers are weakest.

In 2011, a pair of scientists at Michigan State University investigated that issue in a study of 223 students at their university who had filled out a "Morningness Eveningness" questionnaire to determine whether they fit the morning- or evening-person criteria. The subjects were asked—at random—to participate in the experiment either between 8:30 and 9:30 in the morning or between 4:00 and 5:30 in the afternoon. In other words, at the time they were tested, some were at their best, and others at their worst.

Each student was given paper, a pencil, and six problems to solve. They were allotted four minutes for each. Three of the problems were riddles analogous to those cited in chapter 5, such as the one about

Marsha and Marjorie, girls born on the same day to the same mother and the same father, and yet not twins. Finding the solution to these riddles required the subjects to engage in restructuring of their original framework of thought. In the case of Marsha and Marjorie, that means abandoning the picture of two girls that is suggested by the wording of the riddle, for the solution is that Marsha and Marjorie are triplets. The other three were straightforward "analytical" problems, the type that may require careful concentration but can be solved systematically and don't demand elastic thinking. For example: "Bob's father is three times as old as Bob. Four years ago, Bob's father was four times older than Bob. How old are Bob and his father?"

While the students tested at their peak times solved more of the analytical problems, more of the riddles were solved by students tested in their off-peak hours, when their prefrontal cortex was not operating at full capacity. Fatigued people's "more diffuse attentional focus," the researchers wrote, led them to "widening their search through their knowledge network." That widening leads to better performance in problem-solving that requires elastic thinking.

That's good information for those who, in the morning, find themselves in a mental haze or those who, at the end of the day, feel fried and unable to concentrate. For me, it explains a lot. I'm a "night person." I do science best late in the day, whereas in my morning stupor I have done things like crack an egg into the sink and then start to fry the shell in the skillet. And yet I noticed long ago that I'm more successful at *writing* during that foggy and otherwise useless morning time.

I now understand why. Though success in science requires original ideas, once you have an idea, it takes quite a long time to work out its consequences, and it is in that analytical mode that you spend most of your time—hence my success doing science at night. By contrast, when I write, the need for elastic thinking is almost constant. As a result, my morning executive brain "disability" is an advantage in my writing. And so I've learned to listen to my rhythms—that some activities are best done when I still have sleep in my eyes, and others after the weight of the day has painted dark circles beneath them.

Don't Worry, Be Happy

On September 22, 1930, the mother superior of the North American Sisters of Milwaukee, Wisconsin, sent a letter to young nuns in different parts of the country, requesting that they write three-hundred-word essays about their lives. Mostly in their early twenties, the nuns were asked to include outstanding and edifying events from their childhood, and influences that led them to the religious life. The handwritten essays not only contained an accounting of information and feelings; they also reflected, in how they were written, each nun's personality.

The essays were eventually filed away, and they sat untouched for decades. Then, sixty years after they were written, they were stumbled upon by a trio of longevity researchers from the University of Kentucky whose work focused on retired nuns. Amazingly, 180 of the essay writers were among their own research subjects.

Sensing an extraordinary opportunity, the scientists analyzed the essays' emotional content, classifying each as positive, negative, or neutral. And then, over the next nine years of their study, they tabulated the correlation between the nuns' personal disposition and their lifespan. Their conclusion was astonishing: The nuns who'd been the most positive lived about ten years longer than those who'd been the least.

The nun study helped fuel a new field called "positive psychology." Unlike much of psychology, which focuses on people's problems and mental illness, positive psychology focuses on enhancing positive feelings. It is about how you play to the strengths that help you thrive. It's an approach that has become popular with Fortune 500 companies, because research shows that a happy workforce is more productive and creative. That brings us to another way you can relax your cognitive filters without resorting to drugs or technology: by simply improving your mood.

To understand how that works, consider how positive and negative emotions differ. Negative emotions like fear, anger, sadness, and disgust elicit responses in our autonomic nervous system, such as an elevated heart rate or vomiting. Those autonomic reactions reflect the evolutionary purpose of negative emotions. Each is associated with an urge to act in some particular way.* They mean that something is wrong. In prehistoric times, they meant that some sort of danger lurked, and that we needed to take action. Anger encourages us to attack, fear propels us to flee, disgust makes us spit out whatever we have ingested. By contrast, there is no autonomic reaction that distinguishes among different positive emotions. And there is no specific urge that results from happiness, no automatic reaction to serenity, no reflexive response to gratitude.

Because negative emotion creates an instant focus on some particular behavioral response, it narrows the scope of possibilities that your cognitive filters allow through. As a result, a bad mood discourages elastic thinking. For example, in one experiment, negative emotions were induced by having subjects watch film clips of tragic situations. That created an analytic mindset, causing them to perform poorly in a challenge to produce novel word associations.

Good moods are different. Since the positive emotions don't come with action items, they don't narrow your attention. What *do* they do? University of Michigan psychologist Barbara Fredrickson suggested that the purpose of positive emotions is to do precisely the opposite.

Positive emotions, Fredrickson argued, prompt us to consider a wider range of thoughts and actions than are typical. They encourage us to create new relationships, expand our support network, explore our environment, and open ourselves to absorbing information. Those activities increase resilience and lower stress, which is why a happy disposition contributes to survival and longevity.

In order for our brain to accomplish that broadening of attention, Fredrickson reasoned, it must expand the realm of possibilities that our cognitive filters allow through—and that allows us to consider a

---

* In the modern, "civilized" world, there may not actually be an action to take in response to a negative emotion. For example, you might feel anger because another driver rudely cut you off or honked at you, but the best reaction is to do nothing. In such situations, the fact that no response is called for can be unsettling, for your brain is built to produce one. It reflexively prepares to respond, and if you don't, the frustration and feeling of powerlessness that result can be difficult to manage.

broader range of solutions when we encounter problems. Experiments have supported her theory. Positive mood, they reveal, has an effect that is similar to getting high, enabling more original ideas to surface in our conscious minds.

In one study, volunteers who were put into a good mood by watching a funny video or grazing on tasty refreshments did significantly better in tests of elastic thinking than a control group who spent the same period engaged in a mood-neutral activity. As it turns out, the converse is also true: Studies show that successfully applying elastic thinking to solve a problem stimulates your reward circuits and boosts your mood. The result is a virtuous circle in which positive mood and creative problem-solving reinforce each other.

It's good to know the effect of a positive mood on our brains, but even more important is that positive psychology offers ways to achieve that. Its lessons are obviously useful in life regardless of our desire to nurture elastic thinking.

Some of the directives are self-evident, though we don't follow them as often as we should. For example, we'd all benefit from engaging in pleasurable activities, even those as simple as reading a novel or taking a hot bath. Or taking time out to dwell on and celebrate good news, or to share with our friends their good news.

The most famous activity that positive psychologists promote is the "gratitude exercise," in which people are instructed to write down, regularly, three things for which they are grateful. These might be anything from a sunny day to good news about your health. Another intervention capitalizes on research into the satisfaction we derive from doing something for others. For example, on average, it cheers us more to spend money on someone else than on ourselves. That intervention, called the "kindness exercise," is identical to the gratitude exercise, except that you keep a tally of the nice things you have done for people. There has been research on other "listing" exercises as well. In each case, the key to their efficacy seems to be that they make you mindful of positive information about yourself.

And then there is the defensive approach—advice on how to banish cycles of negative thoughts that can invade our minds. The first step is to acknowledge a bad thought and to accept it without immediately attempting to suppress it—acceptance tends to lessen the impact. Next, imagine that it is not you but a friend who is having the thought. What advice would you give that person? If that person made an error at work, for example, you might point out the person's overall positive track record, and that it is unreasonable to expect that you never make a mistake. Now focus on how that advice might apply to you. This defensive approach is powerful—it has even been found to help with the symptoms of depression.

Of all the principles regarding how we open our minds to insight and discovery, to me the best realization is that not only is happiness an end in itself, but it can be a strategy for mental productivity. For those of us who live life focused on what we need to get done rather than on what we need to make us feel good, it's nice to have a reason to add fostering a positive mood to our busy agenda.

Where There's a Will

A few years ago, my mother, who lived in a small house next door to me, needed a new blender. She was then in her late eighties. I told her I would pick one up for her or take her to Best Buy to get one. "No, it's too much trouble," she said. "I don't want to bother you." That was her answer to everything. If I said I was going to the grocery store, where I'd be spending $300 on a carload of food, she'd decline my offer to bring her back a quart of skim milk. She'd say I'd have too much to carry, as if I could handle the other fourteen bags of food, but the added carton would give me a hernia.

The truth was, she took pride in being independent. She walked the mile to the grocery store almost every day and saw an offer of help as an accusation of inadequacy. But Best Buy wasn't the grocery store. It would require a bus ride, and her arthritic legs had trouble getting up and down the steps. I thought about it for a moment, and then I had an idea. "You can buy it online," I told her. "Come over and I'll show you how. You can order it yourself."

My mother was a woman who had never used a computer, and she read large-print books with a magnifying glass. But she agreed. After much effort to find the absolutely cheapest offer, the purchase went relatively smoothly. I didn't tell her that they'd be adding a shipping charge.

A few days later, I popped in and saw the blender sitting on her kitchen counter. I smiled and said, "See how easy that was! It's a different world today." But she wasn't smiling. "It's very nice that a blender shows up on my doorstep," she said. "What's not so nice is that the blender doesn't work. So how do I get my money back? This new world is giving me heartburn."

It was true: The blender was defective. We went to my house and navigated to the website, but it was difficult to find a clear refund policy. It seemed that not only would the return involve a trip to the post office, but she would have to pay the postage. After we had wasted much time, I apologized for steering her wrong and told her to give up. So much for cheap Internet deals. But she wouldn't have it. "Where there's a will, there's a way," she said.

When I was growing up, that was my mother's favorite expression. "How can I both do my Hebrew school homework *and* study for my math test tomorrow?!" *Where there's a will, there's a way.* "How can I possibly make enough money shoveling snow to go to the movies in *two* hours?!" *Where there's a will, there's a way.*

To my mother, if I'd said I wanted to open a dry-cleaning business on Mars, the fact that I'd be 249 million miles from my nearest customer wouldn't be the issue—it would be whether I was determined enough to do it. It was only when I got older that I realized where that attitude had come from—that she'd *where-there's-a-will-there's-a-way*'d herself into surviving a Nazi slave-labor camp, and into creating a decent life in this country despite having lost everyone she loved and arriving at Ellis Island penniless and friendless.

The next evening, I expected that she was still ruminating on the defective appliance, so I stopped by to talk about it some more. But as I stepped in her back door, I was surprised to see not one but *two* identical blenders now sitting on her counter.

"I went to Best Buy and tried to exchange it, but they wouldn't do it without a receipt," she told me. "So I bought another one. Took me all day. Good thing I don't work anymore." She said it as if she'd just retired that week, but she hadn't worked in twenty-seven years.

My mother appeared satisfied with the outcome. I was surprised at how quickly she'd gotten over losing the money on the broken one. It wasn't like her. When I was growing up, if I tossed out a half-eaten orange, she'd look at me as if I were loading a fireplace with $100 bills. We talked for a while, and then, on my way out, I grabbed the busted blender to throw it in the trash. But she told me to leave it. "Why do you think I bought another one?" she said. "I'm not collecting them." I was confused.

"I told you I'd figure it out," she said. "I'm going to return that broken one tomorrow with the receipt from the one I bought today. So this time they'll offer to exchange it, but I'll ask for a refund instead. And since the first one was so cheap, I'll get more than I paid for it." She smiled as if she'd just won the trifecta at the racetrack, though I estimated her "winnings" at $3.17, minus four bus fares.

I've talked a lot about the applications and triumphs of elastic thinking in business, science, and the arts, but just as important are the little ideas like my mother's, which come to us as we try to simply get through the day. I hope her *where there's a will there's a way* mantra is one of the takeaways of this book.

We face many challenges, and sometimes they seem insurmountable. But the human brain, given time and nurturing, has solved countless such problems. On the day she received the broken blender, within my mother's brain, a degree of neophilia attracted her to exploring her options. A reward system motivated her to think, to keep trying until she devised a way to get her money back. A default network of neurons created the associations that eventually generated her clever scheme, while her executive structures kept her attention focused and her cognitive filters kept her from drowning in a multitude of crazy ideas.

My mother is now ninety-five. A few years ago a fog began to roll in, and it has been thickening over time. It is now difficult for her to generate new ideas or imaginative approaches. Scientists tell us that this is because the connections between neurons waste away, weakening the communication between structures that are meant to work together. As we age, and our neural connections dwindle, that balance of power is shifted, and the harmony disrupted. I tried, in writing this book, to provide some insight into those processes. Not because that's comforting when we or those we love begin to decline, but so that we can make the most of our abilities while we still possess them.

In the preceding pages, I've described how elastic thinking arises. I've provided questionnaires to assess your own proclivities. And I've outlined ways to nurture elastic thinking and overcome the barriers that impede it. Some of the suggestions I've offered will likely work for you, while others may not. There is no *one size fits all* when it comes to the human mind. I've seen Deepak Chopra work on a book amid the noise and bustle of a train station, and on an airplane. Physicist Richard Feynman liked to get ideas and scrawl equations while sipping 7-Up in a topless bar in Pasadena (in the days before the topless bars there gave way to tapas bars). On the other hand, Jim Davis, the creator of the *Garfield* comic strip, told me that he had to isolate himself in a hotel room for four days in order to have the uninterrupted peace of mind he needed to create that concept. Jonathan Franzen works alone in an office at the University of California, Santa Cruz, often under a spell so fragile that it is broken by the fragrance of an Indian professor down the hall heating curry in a microwave. I myself cannot do imaginative work if I feel there is a fixed time at which I must quit. And so, if I start work at 10:00 a.m., knowing that I have to put a meatloaf in the oven at 4:00 p.m., that will ruin my whole workday. Our differences are one of the reasons I've emphasized self-knowledge: Only we ourselves, through being mindful of how we operate, can choose the optimal practices for us to follow.

Survival of the Elastic

My father told me of an incident that occurred when he was working for a while as a supervisor of child slaves laboring in a German munitions factory during World War II. As such, he was both a slave laborer himself and a small cog in the German war machine. What the Germans didn't know was that my father was also a leader in the local anti-Nazi underground.

The children my father supervised cared for the chickens, goats, and other animals at the factory, whose presence I now regret I never asked him to explain. The workers were organized into groups of thirty, and every day at precisely 5:00 a.m., my father had to gather his little ones in the cold for roll call. One day, however, when he looked over his kids, he was confronted with a surprise. He had thirty-one.

My father's eyes alighted on the new but familiar face, a boy perhaps nine years of age whose parents had been taken away and killed weeks earlier. My father thought the boy had been killed, too, but he'd apparently managed to hide out. Until now.

The child appeared to be puzzled. He obviously didn't understand why they had been lined up. He didn't know that they were about to be counted. Nor did he know that the person in authority would not accept a count of thirty-one when the correct answer was thirty.

Before my father could talk to the boy, the Gestapo appeared. The lead officer made his count and turned to my father. "You have one extra," he said.

The boy looked at my father, confused. My father's mind raced to find an explanation for the anomaly, but his mental landscape was bare. An extra child could be shot on the spot. So could he. Or all of them. With the Gestapo, you never knew. The officer stared at my father. Seconds ticked past, but his mind remained blank. The children's lives depended on his imagination, but he had let them down.

And then the fugitive child stepped forward. "I've been sick the last month. In the infirmary." He went on for a while, fluidly spinning a yarn, until the officer grew impatient and cut him off. Then the officer made a note on his clipboard. He said to my father, "Now you have thirty-one," and he walked away.

My father told me this story three decades after it had happened. Yet his eyes still welled up as he told it. "That boy, just a young boy, he was like an adult. He thought so quickly. He created a tale, like Isaac Bashevis Singer or Malamud," my father said, putting his act of imagination in the same league as the creations of those two great Jewish authors. Sometime afterward, everyone at the factory, my father included, was sent to a concentration camp. My father did not know whether the boy survived the camp, but, thanks to his powers of elastic thinking, he at least survived that day.

People often speak of the various ways humans differ from other species. Killing members of our own is not one of them. Many aggressive species, such as wolves and chimpanzees, do that. But human murder *is* different from that practiced by other animals. We are the only species in which the intended victim can concoct a story to save himself. That distinction operates in two directions, both of them possible because of our ability to live in our imaginations. First, we have the ability to create stories, and second, we are susceptible to being convinced by them.

War is a time of disruption. Because it brings rapid change, it requires flexibility and the ability to adapt. In those respects, it is a time much like our own, even in the regions of the world that are at peace. For in recent years, we've seen a technological revolution, an information revolution, and economic, political, and social upheavals. We've witnessed amazing new computer applications, sensational scientific discoveries, and a vast enrichment of our intellectual and cultural capital due to globalization. But we've also faced unprecedented new dilemmas.

As our lives have been flooded with novelty and change, they have become more hectic than ever before, at both home and work. We are barraged by a constant stream of information, and thanks to all of our screens and devices, we are in ceaseless contact with dozens, hundreds, and even thousands of other people, rarely (if ever) enjoying any complete downtime.

To be successful today, we must not only cope with the flood of knowledge and data about the present; we must also be able to anticipate the future, because change happens so rapidly that what works well now will be dated and irrelevant tomorrow. The world today is a moving target.

Our brains are information-processors and problem-solving

machines, and certainly our analytical skills are crucial to meeting the challenges we face. But even more important today is the magic of elastic thinking, which can generate new, often wild ideas. Some will prove useless, while others will culminate in the innovative solutions required for the problems of modern existence. To succeed in life today, we need to hone those adaptive skills.

We are lucky to live in a time in which we have begun to understand so much about how the mind works. By describing each of the systems and processes involved in how your brain generates elastic thinking, I hope I have changed the way you think about thinking. And by describing the ways you can alter that operation, I hope I've given you some tools to take charge of it, because there is much you can do to make yourself a more elastic thinker.

Acknowledgments

Unlike a film, which may have a ten-minute credit roll acknowledging people from caterers to casting directors, a book has only the author(s)'s name on it. Writing is indeed a fundamentally solitary and at times lonely profession. But it is also, at crucial, if sporadic, moments, a group effort. In writing *Elastic* I have of course benefited from the work of the hundreds of brilliant and dedicated scientists whose research I quote. But I have also received much brilliant input from friends and colleagues, both on the ideas expressed within the book and the way I express them. I have tortured some of these people with numerous drafts, or pelted them with queries, yet I never experienced them fighting back or even dodging my texts, emails, or phone calls. They are either masochistic, or generous and loyal. Whichever it is, I'd like to thank them here. My wife, Donna Scott, a first-class editor with a sharp critical eye, provided much love, support, and wisdom. Edward Kastenmeier, my gifted and imaginative editor at Penguin Random House, offered many crucial and profound suggestions, and helped to shape this book from start to finish. His assistant, Stella Tan, also provided valuable suggestions. My skillful agent and friend, Susan Ginsburg, provided enthusiastic support, insightful and honest input, and, as always, spectacular wine to nurture our elastic thinking. Josephine Kals and Andrew Weber at Penguin Random House, Stacy Testa at Writers House, and Whitney Peeling also contributed help and advice. And Jennifer McKnew created wonderful illustrations.

For their helpful input, I would also like to thank Ralph Adolphs, Tom Benton, Todd Brun, Antonio Damasio, Zach Halem, Keith Holyoak, Christof Koch, John Kounios, Tom Lyon, Alexei Mlodinow, Nicolai Mlodinow, Olivia Mlodinow, Charles Nicolet, Stanley Oropesa, Sanford Perliss, Marc Raichle, Beth Rashbaum, Randy Rogel, Myron Scholes, Jonathan Schooler, Karen Waltuck, and my remarkable copy editor, Will Palmer. Finally, I am grateful to those whom I had the pleasure of interviewing: Ralph Adolphs, Nancy Andreasen, Mark Beeman, Judy Blume, Antonio Damasio, Jim Davis, Jean Feiwel, Jonathan Franzen, Sidney Harris, Bill T. Jones, John Kounios, Nathan Myhrvold, Stanley McChrystal, Seth McFarlane, Rachel Moore, David Petraeus, and James Warner. Their kind cooperation provided me much insight and added a great deal to the story in this book.

Notes

Introduction

3   On July 6, 2016: The story is pieced together from: Randy Nelson, "Mobile Users Are Spending More Time in Pokémon GO Than Facebook," July 12, 2016, https://sensortower.com/blog/pokemon-go-usage-data; Randy Nelson, "Sensor Tower's Mobile Gaming Leaders for April 2016," May 9, 2016, https://sensortower.com/blog/top-mobile-games-april-2016; Andrew Griffin, "Pokémon Go Beats Porn on Google as Game Becomes Easily One of the Most Popular Ever," July 13, 2006, http://www.independent.co.uk/life-style/gadgets-and-tech/news/pokemon-go-porn-pornography-google-netherlands-uk-canada-a7134136.html; Marcella Machado, "Pokémon Go: Top 10 Records," July 21, 2016, http://www.chupamobile.com/blog/2016/07/21/pokemon-go-top-10-records; Brian Barrett, "Pokemon Go Is Doing Just Fine," *Wired*, September 18, 2016; Sarah Needleman, " 'Pokémon Go' Adds Starbucks Stores as Gyms and PokéStops," *Wall Street Journal*, December 8, 2016, http://www.wsj.com/articles/pokemon-go-adds-starbucks-stores-as-gyms-and-pokestops-1481224993; and Erik Cain, " 'Pokemon Sun' and 'Pokemon Moon' Just Broke a Major Sales Record," *Forbes*, November 30, 2016.

4   Despite all the innovation: Andrew McMillen, "Ingress: The Friendliest Turf War on Earth," https://www.cnet.com/news/ingres-the-friendliest-turf-war-on-earth/February 17, 2015.

5   In 1958, the average life span: Geoff Colvin, "Why Every Aspect of Your Business Is About to Change," *Fortune*, October 22, 2015.

5   Today we consume: John Tierney, "What's New? Exuberance for Novelty Has Benefits," *New York Times*, February 13, 2012.

7   The nematode either solves: J. G. White et al., "The Structure of the Nervous System of the Nematode *Caenorhabditis elegans*: The Mind of a Worm." *Philosophical Transactions of the Royal Society B* 314 (1986): 1–340.

8   It crawls into the gut: Carola Petersen et al., "Travelling at a Slug's Pace: Possible Invertebrate Vectors of *Caenorhabditis* Nematodes," *BMC Ecology* 15, no. 19 (2015).

8   Consider the brooding goose: Temple Grandin and Mark J. Deesing, *Behavioral Genetics and Animal Science* (San Diego: Academic Press, 1998), chapter 1.

1 The Joy of Change

15 In the early days of television: "To Serve Man (*The Twilight Zone*)," *Wikipedia,* https://en.wikipedia.org/wiki/To_Serve_Man_(The_Twilight_Zone).

16 For example, dogs like to explore: Claudia Mettke-Hofmann et al., "The Significance of Ecological Factors for Exploration and Neophobia in Parrots," *Ethology* 108 (2002): 249–72; Patricia Kaulfuss and Daniel S. Mills, "Neophilia in Domestic Dogs (*Canis familiaris*) and Its Implication for Studies of Dog Cognition," *Animal Cognition* 11 (2008): 553–56; Steven R. Lindsay, *Handbook of Applied Dog Behavior and Training,* vol. 1: *Adaptation and Learning* (Ames: Iowa State University Press, 2000). To learn more about the evolution of the domestic dog, see J. Clutton-Brock, "Origins of the Dog: Domestication and Early History," in *The Domestic Dog: Its Evolution, Behaviour, and Interactions with People,* ed. J. Serpell (Cambridge: Cambridge University Press, 1995); Carles Vilà, Peter Savolainen, et al., "Multiple and Ancient Origins of the Domestic Dog," *Science* 276, no. 5319 (June 13, 1997): 1687–89.

17 As a result, in the decade: Mark Ware and Michael Mabe, *The STM Report: An Overview of Scientific and Scholarly Journal Publishing* (The Hague, Netherlands: International Association of Scientific, Technical and Medical Publishers, 2015); Bo-Christer Björk et al., "Scientific Journal Publishing: Yearly Volume and Open Access Availability," *Information Research: An International Electronic Journal* 14, no. 1 (2009); and Richard Van Noorden, "Global Scientific Output Doubles Every Nine Years," *Nature NewsBlog,* May 7, 2014.

18 It wasn't until the 1990s: Andre Infante, "The Evolution of Touchscreen Technology," July 31, 2014, http://www.makeuseof.com/tag/evolution-touchscreen-technology.

18 the academic business literature: The quotes that follow are from Julie Battilana and Tiziana Casciaro, "The Network Secrets of Change Agents," *Harvard Business Review,* July–August 2013, 1; and David A. Garvin and Michael A. Roberto, "Change Through Persuasion," *Harvard Business Review,* February 2005, 26.

20 Because tedium used to be the norm: Patricia Meyer Spacks, *Boredom: The Literary History of a State of Mind* (Chicago: University of Chicago Press, 1995), 13.

21 "Other animals don't do this": David Dobbs, "Restless Genes," *National Geographic,* January 2013.

21 What seems to have changed: Donald C. Johanson, *Lucy's Legacy* (New York: Three Rivers Press, 2009), 267; Winifred Gallagher, *New: Understanding Our Need for Novelty and Change* (New York: Penguin Press, 2012), 18–25.

22 But then, as fossils discovered in China: See, for example, Luca Pagani et al., "Tracing the Route of Modern Humans Out of Africa by Using 225 Human Genome Sequences from Ethiopians and Egyptians," *American Journal of Human Genetics* 96 (2015): 986–91; Huw S. Groucutt et al., "Rethinking the Dispersal of *Homo sapiens* Out of Africa," *Evolutionary Anthropology: Issues, News, and Reviews* 24 (2015): 149–64; Hugo Reyes-Centeno et al., "Genomic and Cranial Phenotype Data Support Multiple Modern Human Dispersals from Africa and a Southern Route into Asia," *Proceedings of the National Academy of Sciences* 111 (2014): 7248–53.

22 It is called DRD4: Richard P. Ebstein et al., "Dopamine D4 Receptor (D4DR) Exon III Polymorphism Associated with the Human Personality Trait of Novelty Seeking," *Nature Genetics* 12 (1996): 78–80.

23 The geographic connection was made: L. J. Matthews and P. M. Butler, "Novelty-Seeking DRD4 Polymorphisms Are Associated with Human Migration Distance Out-of-Africa After Controlling for Neutral Population Gene Structure," *American Journal of Physical Anthropology* 145 (2011): 382–89; and Chuansheng Chen et al., "Population Migration and the Variation of Dopamine D4 Receptor (DRD4) Allele Frequencies Around the Globe," *Evolution and Human Behavior* 20 (1999): 309–24.

23 Those papers reported: Matthews and Butler, "Novelty-Seeking DRD4 Polymorphisms."

24 You may remember that: Ned Zeman, "The Man Who Loved Grizzlies," *Vanity Fair*, October 2, 2009.

25 Below is one of them: From Rick H. Hoyle et al., "Reliability and Validity of a Brief Measure of Sensation Seeking," *Personality and Individual Differences* 32 (2002): 401–14. The scale was actually designed to measure sensation-seeking, defined as the "tendency to seek varied novel complex and intense sensations and experience to take risk for the sake of such experience," but that is highly correlated with novelty-seeking. See: W. F. McCourt et al., "Sensation Seeking and Novelty Seeking: Are They the Same?" *Journal of Nervous Mental Disorders* 181 (May 1993): 309–12.

27 Had I taken a neophilia test: For ages 17–75, see Peter Eachus, "Using the Brief Sensation Seeking Scale (BSSS) to Predict Holiday Preferences," *Personality and Individual Differences* 36 (2004): 141–53. For ages 18–26, see Richard Charnigo et al., "Sensation Seeking and Impulsivity: Combined Associations with Risky Sexual Behavior in a Large Sample of Young Adults," *Journal of Sex Research* 50 (2013): 480–88. For ages 13–17, see Rick H. Hoyle et al., "Reliability and Validity of a Brief Measure of Sensation Seeking," *Personality and Individual Differences* 32 (2002): 401–14.

2 What Is Thought?

31 Anne Green, a "fat, fleshy woman": Carl Zimmer, *Soul Made Flesh* (New York: Atria, 2005), 108–10.

33 The philosopher Karl Popper: Karl Popper, *All Life Is Problem Solving* (Abingdon, UK: Routledge, 2001), 100.

34 And if that food: Toshiyuki Nakagaki et al., "Intelligence: Maze-Solving by an Amoeboid Organism," *Nature* 407 (September 28, 2000): 470.

34 According to the dictionary: "Thinking," Dictionary.com, http://www.dictionary.com/browse/thinking.

34 A textbook on neuroscience: Bryan Kolb and Ian Whishaw, *Introduction to Brains and Behavior* (New York: Worth, 2006), 527.

35 In the late 1970s: Ellen J. Langer et al., "The Mindlessness of Ostensibly Thoughtful Action: The Role of 'Placebic' Information in Interpersonal Interaction," *Journal of Personality and Social Psychology* 36 (1978): 635–42.

36 For instance, relationship researchers: Andrew Christensen and Christopher L. Heavey, "Gender and Social Structure in the Demand/Withdraw Pat-

tern of Marital Conflict," *Journal of Personality and Social Psychology* 59 (1990): 73.

37 William James said: William James, *Memories and Studies* (1911; repr., New York: Longmans, Green, 1924), 237.

37 Luckily, a lot of recent research: See, for example, Amishi P. Jha et al., "Mindfulness Training Modifies Subsystems of Attention," *Cognitive, Affective, & Behavioral Neuroscience* 7 (2007), 109–19; James Carmody and Ruth A. Baer, "Relationships Between Mindfulness Practice and Levels of Mindfulness, Medical and Psychological Symptoms and Well-Being in a Mindfulness-Based Stress Reduction Program," *Journal of Behavioral Medicine* 31 (2008), 23–33.

39 In that address he asked: George Boole, *The Claims of Science*, vol. 15 (Oxford, UK: Oxford University Press, 1851), 15–16.

40 His wife, following the dictates: Stephen Hawking, *God Created the Integers* (Philadelphia: Running Press, 2005), 669–75.

41 A friend of Babbage's: Douglas Hofstadter, *Gödel, Escher, Bach* (New York: Vintage, 1979), 25.

41 Still, Lady Lovelace appreciated: Margaret A. Boden, *The Creative Mind: Myths and Mechanisms* (London: Routledge, 2004), 16.

43 In the words of Andrew Moore: "Artificial Intelligence," *60 Minutes*, October 9, 2016, http://www.cbsnews.com/news/60-minutes-artificial-intelligence-charlie-rose-robot-sophia.

43 There are classical pieces: M. A. Boden, "Creativity and Artificial Intelligence," *Artificial Intelligence* 103 (1998): 347–56. The Brian Eno app is called Bloom.

44 Eno has speculated that: Randy Kennedy, "A New Year's Gift from Brian Eno: A Growing Musical Garden," *New York Times*, January 2, 2017.

44 Consider the following paragraph: Michael Gazzaniga et al., *Cognitive Neuroscience: The Biology of the Mind,* 4th ed. (New York: W. W. Norton, 2014), 74.

45 Tasks that require elastic thinking: David Autor, "Polanyi's Paradox and the Shape of Employment Growth," National Bureau of Economic Research Working Paper No. 20485, 2014.

45 They've built a machine: Quoc Le et al., "Building High-Level Features Using Large Scale Unsupervised Learning," in *Proceedings of the 29th International Conference on Machine Learning,* ed. John Langford and Joelle Pineau (Madison, Wis.: Omnipress, 2012), 81–88.

46 Sanford Perliss, a well-known defense attorney: Told by Sanford Perliss in the keynote lecture, 2017 Perliss Law Symposium on Criminal Trial Practice, April 1, 2017.

3 Why We Think

48 Pat Darcy was forty-one: Eugénie Lhommée et al., "Dopamine and the Biology of Creativity: Lessons from Parkinson's Disease," *Frontiers in Neurology* 5 (2014): 1–11.

50 Kurt Vonnegut wrote: Kurt Vonnegut, *If This Isn't Nice, What Is?* (New York: Rosetta, 2013), 111.

51 As one neuroscientist put it: Nancy Andreasen, "Secrets of the Creative Brain," *The Atlantic*, July–August 2014.

52 We have insight into that question: The material on EVR is from Paul J. Eslinger and Antonio R. Damasio, "Severe Disturbance of Higher Cognition After Bilateral Frontal Lobe Ablation: Patient EVR," *Neurology* 35 (1985): 1731–37; Antonio Damasio, *Descartes' Error: Emotion, Reason, and the Human Brain* (New York: Avon, 1994), 34–51; and Ralph Adolphs, interviewed by author, November 10, 2015. Adolphs is one of the scientists who studied EVR.

53 Without it, EVR could not experience: Wilhelm Hofmann and Loran F. Nordgren, eds., *The Psychology of Desire* (New York: Guilford, 2015), 140.

55 Research suggests: Kimberly D. Elsbach and Andrew Hargadon, "Enhancing Creativity Through 'Mindless' Work: A Framework of Workday Design," *Organization Science* 17 (2006): 470–83.

55 William James expressed the danger: William James, *The Principles of Psychology*, vol. 1 (New York: Henry Holt, 1890), 122.

55 As Swarthmore psychologist Barry Schwartz documented: Barry Schwartz, *The Paradox of Choice: Why More Is Less* (New York: Ecco, 2004); Barry Schwartz et al., "Maximizing Versus Satisficing: Happiness Is a Matter of Choice," *Journal of Personality and Social Psychology* 83 (2002): 1178.

57 One day, Milner's supervisor: Peter Milner, "Peter M. Milner," Society for Neuroscience, https://www.sfn.org/~/media/SfN/Documents/TheHistoryof Neuroscience/Volume%208/PeterMilner.ashx.

57 the nucleus accumbens: R. C. Malenka et al., eds., *Molecular Neuropharmacology: A Foundation for Clinical Neuroscience*, 2nd ed. (New York: McGraw-Hill Medical, 2009), 147–48, 367, 376. To be technically correct, the current leading hypothesis is that the dopamine response is actually caused by "prediction error," the difference between the obtained reward and the expected reward. See Michael Gazzaniga et al., *Cognitive Neuroscience: The Biology of the Mind* (New York: W. W. Norton, 2014), 526–27.

62 One million four hundred thousand: S. Mithen, *The Prehistory of the Mind: The Cognitive Origins of Art and Science* (London: Thames and Hudson, 1996); Marek Kohn and Steven Mithen, "Handaxes: Products of Sexual Selection," *Antiquity* 73 (1999): 518–26.

62 Consider, for example, how the great Russian: Teresa M. Amabile, Beth A. Hennessey, and Barbara S. Grossman, "Social Influences on Creativity: The Effects of Contracted-for Reward," *Journal of Personality and Social Psychology* 50 (1986): 14–23.

63 Many recent studies: Indre V. Viskontas and Bruce L. Miller, "Art and Dementia: How Degeneration of Some Brain Regions Can Lead to New Creative Impulses," in *The Neuroscience of Creativity*, ed. Oshin Vartanian et al. (Cambridge, Mass.: MIT Press, 2013), 126.

63 Difficulty in original thinking: Amabile, "Social Influences on Creativity," 14–23.

63 Young male zebra finches: Kendra S. Knudsen et al., "Animal Creativity: Cross-Species Studies of Cognition," in *Animal Creativity and Innovation*, ed. Alison B. Kaufman and James C. Kaufman (New York: Academic Press, 2015), 213–40.

63 Might artistic talent: Geoffrey Miller, "Mental Traits as Fitness Indicators:

Expanding Evolutionary Psychology's Adaptationism," *Annals of the New York Academy of Sciences* 907 (2000): 62–74.

63 Evolutionary psychologists Martie Haselton and Geoffrey Miller: Martie G. Haselton and Geoffrey F. Miller, "Women's Fertility Across the Cycle Increases the Short-Term Attractiveness of Creative Intelligence," *Human Nature* 17 (2006): 50–73.

65 The children described: Bonnie Cramond, "The Relationship Between Attention-Deficit Hyperactivity Disorder and Creativity," paper presented at the April 1994 meeting of the American Educational Research Association, New Orleans, La., http://files.eric.ed.gov/fulltext/ED371495.pdf.

65 mild brain damage: George Bush, "Attention-Deficit/Hyperactivity Disorder and Attention Networks," *Neuropsychopharmacology* 35 (2010): 278–300.

65 But the most critical traits: N. D. Volkow et al., "Motivation Deficit in ADHD Is Associated with Dysfunction of the Dopamine Reward Pathway," *Molecular Psychiatry* 16 (2011): 1147–54.

67 The theory was tested: Dan T. A. Eisenberg et al., "Dopamine Receptor Genetic Polymorphisms and Body Composition in Undernourished Pastoralists: An Exploration of Nutrition Indices Among Nomadic and Recently Settled Ariaal Men of Northern Kenya," *BMC Evolutionary Biology* 8 (2008): 173–84.

67 Occupational research theorist: Michael Kirton, "Adaptors and Innovators: A Description and Measure," *Journal of Applied Psychology* 61 (1976): 622–45; Michael Kirton, "Adaptors and Innovators: Problem-Solvers in Organizations," in *Readings in Innovation*, ed. David A. Hills and Stanley S. Gryskiewicz (Greensboro, N.C.: Center for Creative Leadership, 1992), 45–66.

68 Thus was born one: Dorothy Leonard and Jeffrey Rayport, "Spark Innovation Through Empathetic Design," *Harvard Business Review on Breakthrough Thinking* (1999): 40.

4 The World Inside Your Brain

72 Though he knew nothing: Rodrigo Quian Quiroga, "Concept Cells: The Building Blocks of Declarative Memory Functions," *Nature Reviews: Neuroscience* 12 (August 2012), 587–94.

73 For example, in 1997, IBM: Shay Bushinsky, "Deus Ex Machina—a Higher Creative Species in the Game of Chess," *AI Magazine* 30, no. 3 (Fall 2009): 63–70.

74 Deep Blue was far faster: Robert Weisberg, *Creativity* (New York: John Wiley and Sons, 2006), 38.

74 Meanwhile, processors have become: Bushinsky, "Deus Ex Machina," 63–70.

74 They are excellent at learning: Cade Metz, "In a Huge Breakthrough, Google's AI Beats a Top Player at the Game of Go," *Wired*, January 27, 2016.

76 Whatever it includes for you: Derek C. Penn et al., "Darwin's Mistake: Explaining the Discontinuity Between Human and Nonhuman Minds," *Behavioral and Brain Sciences* 31 (2008): 109–20.

78 Neuroscientists call the neurons: Charles E. Connor, "Neuroscience: Friends and Grandmothers," *Nature* 435 (2005): 1036–37.

78 We are capable of encoding: Quiroga, "Concept Cells," 587–94.

78 The fact that neurons: L. Gabora and A. Ranjan, "How Insight Emerges," in *The Neuroscience of Creativity,* ed. Oshin Vartanian et al. (Cambridge, Mass.: MIT Press, 2013), 19–43.

79 If the net input: Bryan Kolb and Ian Whishaw, *Introduction to Brains and Behavior* (New York: Worth, 2006), 45, 76–81, 157.

81 Army ants organize: Hasan Guclu, "Collective Intelligence in Ant Colonies," *The Fountain* 48 (October–December 2004).

81 "Ants never make more ants": Deborah Gordon, "The Emergent Genius of Ant Colonies," TED Talk, February 2003, http://www.ted.com/talks/deborah _gordon_digs_ants.

85 He proudly showed me: Nathan Myhrvold, interviewed by author, January 15, 2016.

5 The Power of Your Point of View

91 David Wallerstein was not someone: Greg Critser, *Fat Land: How Americans Became the Fattest People in the World* (New York: Houghton Mifflin, 2004), 20–29.

95 For, though the company: Geoff Colvin, "Why Every Aspect of Your Business Is About to Change," *Fortune,* October 22, 2015.

96 And it is looking forward: Michal Addady, "Nike Exec Says We'll Be 3D Printing Sneakers at Home Soon," *Fortune,* October 7, 2015.

96 Consider these brainteasers: Vinod Goel et al., "Differential Modulation of Performance in Insight and Divergent Thinking Tasks with tDCS," *Journal of Problem Solving* 8 (2015): 2.

99 Computer scientist Douglas Hofstadter: Douglas Hofstadter, *Gödel, Escher, Bach* (New York: Vintage, 1979), 611–13.

101 When presented with this problem: Robert Weisberg, *Creativity* (New York: John Wiley and Sons, 2006), 306–7.

102 And so Bombelli questioned: Edna Kramer, *The Nature and Growth of Modern Mathematics* (Princeton, N.J.: Princeton University Press, 1983), 70.

104 Whatever its source: Shinobu Kitayama and Ayse K. Uskul, "Culture, Mind, and the Brain: Current Evidence and Future Directions," *Annual Review of Psychology* 62 (2011): 419–49; Shinobu Kitayama et al., "Perceiving an Object and Its Context in Different Cultures: A Cultural Look at New Look," *Psychological Science* 14 (May 2003): 201–6.

105 It ranks the United States: Scott Shane, "Why Do Some Societies Invent More Than Others?" Working Paper Series 8/90, Wharton School, September 1990. Some countries were excluded due to the unavailability of data in certain years.

106 In this study: "A New Ranking of the World's Most Innovative Countries," Economist Intelligence Unit report, April 2009, http://graphics.eiu.com/PDF /Cisco_Innovation_Complete.pdf.

107 On the other hand, exposing: Karen Leggett Dugosh and Paul B. Paulus, "Cognitive and Social Comparison in Brainstorming," *Journal of Experimental Social Psychology* 41 (2005): 313–20; and Karen Leggett Dugosh et al., "Cognitive Stimulation in Brainstorming," *Journal of Personality and Social Psychology* 79 (2005): 722–35.

6 Thinking When You're Not Thinking

108   Lying in her bed: The story of the creation of *Frankenstein* is from Frank Barron, et al., eds., *Creators on Creating: Awakening and Cultivating the Imaginative Mind* (New York: Tarcher/Penguin, 1997), 91–95.

110   Known as the brain's *default mode:* Marcus Raichle et al., "Rat Brains Also Have a Default Network," *Proceedings of the National Academy of Sciences* 109 (March 6, 2012): 3979–84.

110   There is currently an explosion: For Raichle's seminal work, see Marcus E. Raichle et al., "A Default Mode of Brain Function," *Proceedings of the National Academy of Sciences* 98 (2001): 676–82. The history of the research is discussed in Randy L. Buckner et al., "The Brain's Default Network," *Annals of the New York Academy of Sciences* 1124 (2008): 1–38.

111   The tale begins in 1897: For Berger's story, see David Millett, "Hans Berger: From Psychic Energy to the EEG," *Perspectives in Biology and Medicine* 44 (Autumn 2001): 522–42; T. J. La Vaque, "The History of EEG: Hans Berger, Psychophysiologist; A Historical Vignette," *Journal of Neurotherapy* 3 (Spring 1999): 1–9; and P. Gloor, "Hans Berger on the Electroencephalogram of Man," *EEG Clinical Neurophysiology* 28 (Suppl. 1969): 1–36.

111   One said he was: La Vaque, "The History of EEG," 1–2.

111   Another, who would later serve: Millett, "Hans Berger," 524.

112   As he wrote many years later: La Vaque, "The History of EEG," 1–2.

114   He preached about the importance: See the account in Marcus Raichle, "The Brain's Dark Energy," *Scientific American*, March 2010, 46; and Millett, "Hans Berger," 542. But there were some exceptions, especially in Britain, for instance; E. D. Adrian and B. H. C. Matthews, "Berger Rhythm: Potential Changes from the Occipital Lobes in Man," *Brain* 57 (1934): 355–85.

115   In May 1941: La Vaque, "The History of EEG," 8.

115   "I would like to draw attention": H. Berger, "Über das Elektrenkephalogramm des Menschen," *Archiv für Psychiatrie und Nervenkrankheiten* 108 (1938): 407. Translation from La Vaque, "The History of EEG," 8.

116   On his study wall: La Vaque, "The History of EEG," 8.

116   Not for her: Nancy Andreasen, interviewed by author, April 10, 2015.

117   "I was the first woman": Nancy Andreasen, "Secrets of the Creative Brain," *The Atlantic*, July–August 2014.

119   However, Andreasen had just scratched: Randy L. Buckner, "The Serendipitous Discovery of the Brain's Default Network," *Neuroimage* 62 (2012): 1137–45.

120   With great pain, scientists can: M. D. Hauser, S. Carey, and L. B. Hauser, "Spontaneous Number Representation in Semi-Free-Ranging Rhesus Monkeys," *Proceedings of the Royal Society of London B* 267 (2000): 829–33.

121   Consider the famous case: Antonio R. Damasio and G. W. Van Hoesen, "Emotional Disturbances Associated with Focal Lesions of the Limbic Frontal Lobe," in *Neuropsychology of Human Emotion*, ed. Kenneth Heilman and Paul Satz (New York: Guilford, 1983), 85–110.

123   Fifty-eight percent of adults: Larry D. Rosen et al., "The Media and Technology Usage and Attitudes Scale: An Empirical Investigation," *Computers in Human Behavior* 29 (2013): 2501–11; and Nancy A. Cheever et al., "Out of Sight Is Not Out of Mind: The Impact of Restricting Wireless Mobile Device

Use on Anxiety Levels Among Low, Moderate and High Users," *Computers in Human Behavior* 37 (2014): 290–97.

124 In one study: Russell B. Clayton et al., "The Extended iSelf: The Impact of iPhone Separation on Cognition, Emotion, and Physiology," *Journal of Computer-Mediated Communication* 20, no. 2 (2015): 119–35.

124 Says David Greenfield: Emily Sohn, "I'm a Smartphone Addict, but I Decided to Detox," *Washington Post*, February 8, 2016.

124 "A massive increase". C. Shawn Green and Daphne Bavelier, "The Cognitive Neuroscience of Video Games," in *Digital Media: Transformations in Human Communication*, ed. Paul Messaris and Lee Humphreys (New York: Peter Lang, 2006), 211–23. See also Shaowen Bao et al., "Cortical Remodelling Induced by Activity of Ventral Tegmental Dopamine Neurons," *Nature* 412 (2001): 79–83.

126 When you get back to join: Marc G. Berman et al., "The Cognitive Benefits of Interacting with Nature," *Psychological Science* 19 (2008): 1207–12.

126 Research shows a positive correlation: Joseph R. Cohen and Joseph R. Ferrari, "Take Some Time to Think This Over: The Relation Between Rumination, Indecision, and Creativity," *Creativity Research Journal* 22 (2010): 68–73.

127 But Leonardo "talked to him extensively": Giorgio Vasari, *The Lives of the Artists* (Oxford, UK: Oxford University Press, 1991), 290.

7 The Origin of Insight

128 One cold day a few weeks later: For the Low story, see Craig Nelson, *The First Heroes: The Extraordinary Story of the Doolittle Raid—America's First World War II Victory* (New York: Penguin, 2003); Carroll V. Glines, *The Doolittle Raid* (Atglen, Pa.: Schiffer Military/Aviation History, 1991), 13; Don M. Tow, "The Doolittle Raid: Mission Impossible and Its Impact on the U.S. and China," http://www.dontow.com/2012/03/the-doolittle-raid-mission-impossible-and-its-impact-on-the-u-s-and-china; and Kirk Johnson, "Raiding Japan on Fumes in 1942, and Surviving to Tell How Fliers Did It," *New York Times*, February 1, 2014.

130 The Japanese fleet was essentially crippled: John Keegan, *The Second World War* (New York: Penguin, 2005), 275.

131 When asked how: Glines, *Doolittle Raid*, 15.

131 Roger Sperry pondered: For Sperry's story, see R. W. Sperry, "Roger W. Sperry Nobel Lecture, 8 December 1981," *Nobel Lectures, Physiology or Medicine* 1990 (1981); Norman Horowitz et al., "Roger Sperry, 1914–1994," *Engineering & Science* (Summer 1994): 31–38; Robert Doty, "Physiological Psychologist Roger Wolcott Sperry 1913–1994," *APS Observer* (July–August 1994): 34–35; and Nicholas Wade, "Roger Sperry, a Nobel Winner for Brain Studies, Dies at 80," *New York Times*, April 20, 1994.

132 Some of Sperry's colleagues: Roger Sperry, Nobel Lecture, Nobelprize.org, December 8, 1981.

132 At first Sperry: R. W. Sperry, "Cerebral Organization and Behavior," *Science* 133 (June 2, 1961): 1749–57.

132 "Unable to perform": Ibid.

133 "Each of the divided": Ibid.

133 That's when Joseph Bogen: Ivan Oransky, "Joseph Bogen," *The Lancet* 365 (2005): 1922.

133 For example, in one case: Deepak Chopra and Leonard Mlodinow, *War of the Worldviews* (New York: Harmony, 2011): 179–80.

138 "The origin of insight": John Kounios, interviewed by author, February 23, 2015.

139 Although they can still speak: Mark Beeman, interviewed by author, February 23, 2015.

141 "It's being reported that": *Conan*, TBS, March 16, 2015.

142 They decided to use: E. M. Bowden and M. J. Beeman, "Getting the Idea Right: Semantic Activation in the Right Hemisphere May Help Solve Insight Problems," *Psychological Science* 9 (1998): 435–40.

143 About 40 percent more puzzles: Mark Jung-Beeman et al., "Neural Activity When People Solve Verbal Problems with Insight," *PLOS Biology* 2 (April 2004): 500–507.

144 One role of the ACC: Simon Moss, "Anterior Cingulate Cortex," Sicotests, http://www.psych-it.com.au/Psychlopedia/article.asp?id=263; Carola Salvi et al., "Sudden Insight Is Associated with Shutting Out Visual Inputs," *Psychonomic Bulletin and Review* 22, no. 6 (December 2015): 1814–19; and John Kounios and Mark Beeman, "The Cognitive Neuroscience of Insight," *Annual Review of Psychology* 65 (2014): 1–23.

147 He asked the man: John Kounios and Mark Beeman, *The Eureka Factor: Aha Moments, Creative Insight, and the Brain* (New York: Random House, 2015), 195–96.

147 A 2012 study: Lorenza S. Colzato et al., "Meditate to Create: The Impact of Focused-Attention and Open-Monitoring Training on Convergent and Divergent Thinking," *Frontiers in Psychology* 3 (2012): 116.

147 If you are interested: Richard Chambers et al., "The Impact of Intensive Mindfulness Training on Attentional Control, Cognitive Style, and Affect," *Cognitive Therapy and Research* 32 (2008): 303–22.

149 For example, research shows: J. Meyers-Levy and R. Zhou, "The Influence of Ceiling Height: The Effect of Priming on the Type of Processing That People Use," *Journal of Consumer Research* 34 (2007): 1741–86.

8 How Thought Freezes Over

154 Subjects are given a box: R. L. Dominowski and P. Dollob, "Insight and Problem Solving," in *The Nature of Insight*, ed. R. J. Sternberg and J. E. Davidson (Cambridge, Mass.: MIT Press, 1995), 33–62.

155 Young children, when given this: Tim P. German and Margaret Anne Defeyter, "Immunity to Functional Fixedness in Children," *Psychonomic Bulletin and Review* 7 (2000): 707–12.

155 In one study, so did: Tim P. German and H. Clark Barrett, "Functional Fixedness in a Technologically Sparse Culture," *Psychological Science* 16b (2005): 1–5.

157 In the preface to his 1936 book: John Maynard Keynes, *General Theory of Employment, Interest and Money* (New York: Harvest/Harcourt, 1936), vii.

158 But, he added, he believed: James Jeans, "A Comparison Between Two Theories of Radiation," *Nature* 72 (July 27, 1905): 293–94.

159   And yet, Arendt noted: Hannah Arendt, "Thinking and Moral Consider-
      ations," *Social Research* 38 (Autumn 1971): 423.

159   Or, as the photographer Dorothea Lange: Milton Meltzer, *Dorothea Lange: A
      Photographer's Life* (Syracuse, N.Y.: Syracuse University Press, 2000), 140.

160   It examined ten years of data: B. Jena Anapam et al., "Mortality and Treatment
      Patterns Among Patients Hospitalized with Acute Cardiovascular Conditions
      During Dates of National Cardiology Meetings," *JAMA Internal Medicine* 10
      (2014): E1–E8.

160   The *JAMA* study didn't pinpoint: Merim Bilalić and Peter McLeod, "Why
      Good Thoughts Block Better Ones," *Scientific American* 310 (January 3,
      2014): 74–79.

160   That alarming finding: Doron Garfinkel, Sarah Zur-Gil, and H. Ben-Israel,
      "The War Against Polypharmacy: A New Cost-Effective Geriatric-Palliative
      Approach for Improving Drug Therapy in Disabled Elderly People," *Israeli
      Medical Association Journal* 9 (2007): 430.

161   The kind of elastic thinking: Erica M. S. Sibinga and Albert W. Wu, "Clinician
      Mindfulness and Patient Safety," *Journal of the American Medical Associa-
      tion* 304 (2010): 2532–33.

161   "In the military": McChrystal quotes are from Stanley McChrystal, inter-
      viewed by author, January 13, 2016.

162   As McChrystal's successor: David Petraeus, interviewed by author, February 16,
      2016.

162   The tale has since become: See, for example, Abraham Rabinovich, *The Yom
      Kippur War: The Epic Encounter That Transformed the Middle East* (New
      York: Schocken Books, 2004); David T. Buckwalter, "The 1973 Arab-Israeli
      War," in *Case Studies in Policy Making & Process*, ed. Shawn W. Burns
      (Newport, R.I.: Naval War College, 2005), 17; and Uri Bar-Joseph and Arie W.
      Kruglanki, "Intelligence Failure and the Need for Cognitive Closure," *Political
      Psychology* 24 (2003): 75–99.

166   "He far exceeded": James Warner, interviewed by author, December 14,
      2015.

166   His legacy, as a *Forbes*: Dan Schwabel, "Stanley McChrystal: What the Army
      Can Teach You About Leadership," *Forbes*, July 13, 2015.

166   They began by showing: Bilalić and McLeod, "Why Good Thoughts Block Bet-
      ter Ones," 74–79; Merim Bilalić et al., "The Mechanism of the Einstellung
      (Set) Effect: A Pervasive Source of Cognitive Bias," *Current Directions in Psy-
      chological Science* 19 (2010): 111–15.

167   The solutions appear: In the board position on the left, the familiar "smothered
      mate" solution is possible: (1) Qe6+ Kh8 (2) Nf7+ Kg8 (3) Nh6++ Kh8 (4)
      Qg8+ Rxg8 (5) Nf7#. The shorter, optimal solution is: (1) Qe6+ Kh8 (2) Qh6
      Rd7 (3) Qxh7#, or (2) . . . Kg8 (3) Qxg7#. In the board position on the right, the
      smothered mate is no longer possible, because black's bishop now covers f7. The
      optimal solution is still possible: (1) Qe6+ Kh8 (If (1) . . . Kf8, 2 Nxh7#) (2) Qh6
      Rd7 (3) Qxh7#, or (2) . . . Kg8 (3) Qxg7#, or (2). . . . Bg6 (3) Qxg7#. The crucial
      squares for the familiar solution are marked by rectangles (f7, g8, and g5), and
      the optimal solution by circles (b2, h6, h7, and g7) in (a). From Bilalić, et al.,
      "Why Good Thoughts Block Better Ones: The Mechanism of the Pernicious
      Einstellung Effect," *Cognition* 108 (2008): 652–61.

168 When psychologists study: Victor Ottati et al., "When Self-Perceptions Increase Closed-Minded Cognition: The Earned Dogmatism Effect," *Journal of Experimental Social Psychology* 61 (2015): 131–38.

169 "Social norms dictate": Ibid.

169 Consider a study, performed about: Serge Moscovici, Elisabeth Lage, and Martine Naffrechoux, "Influence of a Consistent Minority on the Responses of a Majority in a Color Perception Task," *Sociometry* 32, no. 4 (1969): 365–80.

170 Other experiments show that dissent: C. J. Nemeth, "Minority Influence Theory," in *Handbook of Theories of Social Psychology*, ed. P. Van Lange, A. Kruglanski, and T. Higgins (New York: Sage, 2009).

171 Major General Zeira told officers: Uri Bar-Joseph and Arie W. Kruglanki, "Intelligence Failure and the Need for Cognitive Closure," *Political Psychology* 24 (2003): 75–99.

9 Mental Blocks and Idea Filters

173 Our conscious brains can process: Ap Dijksterhuis, "Think Different: The Merits of Unconscious Thought in Preference Development and Decision Making," *Journal of Personality and Social Psychology* 87 (2004): 586–98.

176 Despite its simplicity: T. C. Kershaw and S. Ohlsson, "Multiple Causes of Difficulty in Insight: The Case of the Nine-Dot Problem," *Journal of Experimental Psychology: Learning, Memory, and Cognition* 30 (2004): 3–13; and R. W. Weisberg and J. W. Alba, "An Examination of the Alleged Role of Fixation in the Solution of Several Insight Problems," *Journal of Experimental Psychology: General* 110 (1981): 169–92.

178 With those additions: James N. MacGregor, Thomas C. Ormerod, and Edward P. Chronicle, "Information Processing and Insight: A Process Model of Performance on the Nine-Dot and Related Problems," *Journal of Experimental Psychology: Learning, Memory, and Cognition* 27 (2001): 176.

178 Another way to increase success: Ching-tung Lung and Roger L. Dominowski, "Effects of Strategy Instructions and Practice on Nine-Dot Problem Solving," *Journal of Experimental Psychology: Learning, Memory, and Cognition* 11, no. 4 (January 1985): 804–11.

180 In 2012, over a period of: Richard P. Chi and Allan W. Snyder, "Brain Stimulation Enables the Solution of an Inherently Difficult Problem," *Neuroscience Letters* 515 (2012): 121–24.

180 For example, in one study, researchers: Ibid.

181 When researchers used transcranial: See, for example, Carlo Cerruti and Gottfried Schlaug, "Anodal Transcranial Stimulation of the Prefrontal Cortex Enhances Complex Verbal Associative Thought," *Journal of Cognitive Neuroscience* 21 (October 2009); M. B. Iyer et al., "Safety and Cognitive Effect of Frontal DC Brain Polarization in Healthy Individuals," *Neurology* 64 (March 2005): 872–75; Carlo Reverbi et al., "Better Without (Lateral) Frontal Cortex? Insight Problems Solved by Frontal Patients," *Brain* 128 (2005): 2882–90; and Arthur P. Shimamura, "The Role of the Prefrontal Cortex in Dynamic Filtering," *Psychobiology* 28 (2000): 207–18.

181 Though all mammals: Michael Gazzinga, *Human: The Science Behind What Makes Us Unique* (New York: HarperCollins, 2008), 17–22. The lateral pre-

frontal cortex is a region whose microscopic structure looks distinctive, and in which certain distinct functions are centered, but it doesn't stand out on viewing, like a heart or a kidney. If you were to look at a brain, there is not usually a sharp, recognizable physical delineation.

181 A key part of your "executive brain": Joaquin M. Fuster, "The Prefrontal Cortex—an Update: Time Is of the Essence," *Neuron* 30 (May 2001): 319–33.

181 It is your lateral prefrontal cortex: John Kounios and Mark Beeman, "The Cognitive Neuroscience of Insight," *Annual Reviews in Psychology* 65 (2014): 71–93; E. G. Chrysikou et al., "Noninvasive Transcranial Direct Current Stimulation over the Left Prefrontal Cortex Facilitates Cognitive Flexibility in Tool Use," *Cognitive Neuroscience* 4 (2013): 81–89.

183 Two-time Nobel laureate: Mihaly Csikszentmihalyi, *Creativity: The Psychology of Discovery and Invention* (New York: Harper Perennial, 2013), 116.

184 "I see," Myhrvold said: Nathan Myhrvold, interviewed by author, January 15, 2016.

184 Here's how that discussion: George Lucas et al., "Raiders of the Lost Ark" story conference transcript, January 1978, http://maddogmovies.com/almost/scripts/raidersstoryconference1978.pdf.

185 He has won Emmy awards: The "No. 1 Offender" label appeared in Claire Hoffman, "No. 1 Offender in Hollywood," *New Yorker*, June 18, 2012.

185 "It is hard to maintain a mindset": Seth MacFarlane, interviewed by author, January 29, 2016.

186 That's what fascinates me: Ken Tucker, *Family Guy* review, *Entertainment Weekly*, April 19, 1999, http://www.ew.com/article/1999/04/09/family-guy.

186 As long as those: Nitin Gogtay et al., "Dynamic Mapping of Human Cortical Development During Childhood Through Early Adulthood," *Proceedings of the National Academy of Sciences of the United States of America* 101 (2004): 8174–79.

186 Author and poet Ursula K. Le Guin: Le Guin denies being the source of the quote, and it cannot be found in any of her writings. See her comment on the matter in Ursula K. Le Guin, "A Child Who Survived," blog entry on Book View Café posted on December 28, 2015, http://bookviewcafe.com/blog/2015/12/28/a-child-who-survived/.

10 The Good, the Mad, and the Odd

188 In 1951, the *Proceedings of:* See Matan Shelomi, "Mad Scientist: The Unique Case of a Published Delusion," *Science and Engineering Ethics* 9 (2013): 381–88.

189 One finds a higher-than-average: Shelley Carson, "Creativity and Psychopathology," in *The Neuroscience of Creativity*, ed. Oshin Vartanian et al. (Cambridge, Mass.: MIT Press, 2013), 175–203.

189 And then there was the brilliant: For Tesla's story, see Margaret Cheney, *Tesla: Man Out of Time* (New York: Simon & Schuster, 2011).

190 The first progress toward answers: A. Laguerre, M. Leboyer, and F. Schürhoff, "The Schizotypal Personality Disorder: Historical Origins and Current Status," *L'Encéphale* 34 (2008): 17–22; and Shelley Carson, "The Unleashed Mind," *Scientific American*, May 2011, 22–29.

190 They were not schizophrenic: Leonard L. Heston, "Psychiatric Disorders in

Foster Home Reared Children of Schizophrenic Mothers," *British Journal of Psychiatry* 112 (1966): 819–25.

191   Below is an example: Eduardo Fonseca-Pedrero et al., "Validation of the Schizotypal Personality Questionnaire—Brief Form in Adolescents," *Schizophrenia Research* 111 (2009): 53–60.

192   Over the years, those who scored: See, for example, Bradley S. Folley and Sohee Park, "Verbal Creativity and Schizotypal Personality in Relation to Prefrontal Hemispheric Laterality: A Behavioral and Near-Infrared Optical Imaging Study," *Schizophrenia Research* 80 (2005): 271–82.

192   The eccentric/elastic connection arises: Carson, "Unleashed Mind," 22; Rémi Radel et al., "The Role of (Dis)Inhibition in Creativity: Decreased Inhibition Improves Idea Generation," *Cognition* 134 (2015): 110–20; and Marjaana Lindeman et al., "Is It Just a Brick Wall or a Sign from the Universe? An fMRI Study of Supernatural Believers and Skeptics," *Social Cognitive and Affective Neuroscience* 8 (2012): 943–49, and the studies cited within. Note that this paper refers to the inferior frontal gyrus (IFG) rather than the lateral prefrontal cortex—the ventral aspect of the lateral prefrontal cortex is situated on the IFG.

193   "Because the ideas I had": Carson, "Creativity and Psychopathology," 180–81.

193   Nash was an extreme case: Lindeman, "Is It Just a Brick Wall," and the studies cited within. See also Deborah Kelemen and Evelyn Rosset, "The Human Function Compunction: Teleological Explanation in Adults," *Cognition* 111 (2009): 138–43.

194   In Mozart's own words: Cliff Eisen and Simon P. Keefe, eds., *The Cambridge Mozart Encyclopedia* (Cambridge, UK: Cambridge University Press, 2006), 102.

194   In one study, Geoffrey Wills: Geoffrey I. Wills, "Forty Lives in the Bebop Business: Mental Health in a Group of Eminent Jazz Musicians," *British Journal of Psychiatry* 183 (2003): 255–59.

196   And then there are: For Einstein, see Graham Farmelo, *The Strangest Man: The Hidden Life of Paul Dirac, Mystic of the Atom* (New York: Basic Books, 2009), 344; for Newton, see Leonard Mlodinow, *The Upright Thinkers* (New York: Pantheon, 2015).

196   Those with a higher IQ: Shelley H. Carson, Jordan B. Peterson, and Daniel M. Higgins, "Decreased Latent Inhibition Is Associated with Increased Creative Achievement in High-Functioning Individuals," *Journal of Personality and Social Psychology* 85 (2003): 499.

196   The difficulty of shaping: Except for the bipolar disorder in authors. See Simon Kyaga et al., "Mental Illness, Suicide and Creativity: 40-Year Prospective Total Population Study," *Journal of Psychiatric Research* 47 (2013): 83–90.

196   Growing up in the 1940s: The Judy Blume story is from Judy Blume, interviewed by author, December 2, 2015.

198   Researchers asked subjects to analyze: Vinod Goel et al., "Dissociation of Mechanisms Underlying Syllogistic Reasoning," *Neuroimage* 12 (2000): 504–14.

11 Liberation

200   Some years ago, a scientist: This account was written in 1969 for publication in *Marijuana Reconsidered* (Cambridge, Mass.: Harvard University Press, 1971).

201 One of the few early studies: Charles T. Tart, "Marijuana Intoxication: Common Experiences," *Nature* 226 (May 23, 1970): 701–4.

202 In one of those, a 2012 study: Gráinne Schafer et al., "Investigating the Interaction Between Schizotypy, Divergent Thinking and Cannabis Use," *Consciousness and Cognition* 21 (2012): 292–98.

202 The marijuana had boosted: Ibid.

203 Particularly worrisome is: Kyle S. Minor et al., "Predicting Creativity: The Role of Psychometric Schizotypy and Cannabis Use in Divergent Thinking," *Psychiatry Research* 220 (2014): 205–10.

204 Wilson began using marijuana: Stefano Belli, "A Psychobiographical Analysis of Brian Douglas Wilson: Creativity, Drugs, and Models of Schizophrenic and Affective Disorders," *Personality and Individual Differences* 46 (2009): 809–19.

204 He credited the drug's influence: *Beautiful Dreamer: Brian Wilson and the Story of SMiLE*, directed by David Leaf, produced by Steve Ligerman (Rhino Video, 2004); and Brian Wilson and T. Gold, *Wouldn't It Be Nicer: My Own Story* (New York: Bloomsbury, 1991), 114.

204 In 1982, Wilson was diagnosed: Alexis Petridis, "The Astonishing Genius of Brian Wilson," *The Guardian*, June 24, 2011.

204 For example, in a 2012 study: Andrew F. Jarosz, Gregory J. H. Colflesh, and Jennifer Wiley, "Uncorking the Muse: Alcohol Intoxication Facilitates Creative Problem Solving," *Consciousness and Cognition* 21 (2012): 487–93.

205 A group at Oxford: Robin L. Carhart-Harris et al., "Neural Correlates of the LSD Experience Revealed by Multimodal Neuroimaging," *Proceedings of the National Academy of Sciences* 113 (2016): 4853–58; Robin L. Carhart-Harris et al., "The Entropic Brain: A Theory of Conscious States Informed by Neuroimaging Research with Psychedelic Drugs," *Frontiers in Human Neuroscience* 8 (2014): 1–22.

206 "It was the most intense": Catherine Elsworth, "Isabel Allende: Kith and Tell," *The Telegraph*, March 21, 2008.

207 Ayahuasca seems to function: K. P. C. Kuypers et al., "Ayahuasca Enhances Creative Divergent Thinking While Decreasing Conventional Convergent Thinking," *Psychopharmacology* 233 (2016): 3395–3403; and Joan Francesc Alonso et al., "Serotonergic Psychedelics Temporarily Modify Information Transfer in Humans," *International Journal of Neuropsychopharmacology* 18 (2015): pyvo39.

207 Allende said she faced: Elsworth, "Isabel Allende: Kith and Tell."

208 In 2015, a group of researchers: Rémi Radel et al., "The Role of (Dis)Inhibition in Creativity: Decreased Inhibition Improves Idea Generation," *Cognition* 134 (2015): 110–20.

209 Not everyone regularly feels: Charalambos P. Kyriacou and Michael H. Hastings, "Circadian Clocks: Genes, Sleep, and Cognition," *Trends in Cognitive Science* 14 (2010): 259–67.

209 In 2011, a pair of scientists: Mareike B. Wieth and Rose T. Zacks, "Time of Day Effects on Problem Solving: When the Non-Optimal Is Optimal," *Thinking & Reasoning* 17 (2011): 387–401.

211 On September 22, 1930: Deborah D. Danner, David A. Snowdon, and Wallace V. Friesen, "Positive Emotions in Early Life and Longevity: Findings

from the Nun Study," *Journal of Personality and Social Psychology* 80 (2001): 804.

211 To understand how that works: Barbara L. Fredrickson, "The Value of Positive Emotions," *American Scientist* 91 (2003): 330–35.

212 University of Michigan psychologist: Barbara L. Fredrickson and Christine Branigan, "Positive Emotions Broaden the Scope of Attention and Thought-Action Repertoires," *Cognitive Emotions* 19 (2005): 313–32.

213 Experiments have supported: See the studies in Fredrickson and Branigan, "Positive Emotions Broaden the Scope"; and Soghra Akbari Chermahini and Bernhard Hommel, "Creative Mood Swings: Divergent and Convergent Thinking Affect Mood in Opposite Ways," *Psychological Research* 76 (2012): 634–40.

213 The most famous activity: Joshua Rash et al., "Gratitude and Well-Being: Who Benefits the Most from a Gratitude Intervention?," *Applied Psychology: Health and Well-Being* 3 (2011): 350–69.

213 And then there is the defensive: Justin D. Braun et al., "Therapist Use of Socratic Questioning Predicts Session-to-Session Symptom Change in Cognitive Therapy for Depression," *Behaviour Research and Therapy* 70 (2015): 32–37.

216 Scientists tell us that: See, for example, Cheryl L. Grady et al., "A Multivariate Analysis of Age-Related Differences in Default Mode and Task-Positive Networks Across Multiple Cognitive Domains," *Cerebral Cortex* 20 (2009): 1432–47.

219 Many aggressive species: See, for example, Michael L. Wilson et al., "Lethal Aggression in *Pan* Is Better Explained by Adaptive Strategies Than Human Impacts," *Nature* 513 (2014): 414–17; and Richard W. Wrangham, "Evolution of Coalitionary Killing," *American Journal of Physical Anthropology* 110 (1999): 1–30.

Index

Page numbers in *italics* refer to illustrations.

ABOUT THE AUTHOR

LEONARD MLODINOW received his Ph.D. in theoretical physics from the University of California, Berkeley, was an Alexander von Humboldt Fellow at the Max Planck Institute, and was on the faculty of the California Institute of Technology. His previous books include the best sellers *Subliminal* (winner of the PEN/E. O. Wilson Literary Science Writing Award), *The Drunkard's Walk: How Randomness Rules Our Lives* (a *New York Times* Notable Book), *War of the Worldviews* (with Deepak Chopra), and *The Grand Design* and *A Briefer History of Time* (both with Stephen Hawking), as well as *The Upright Thinkers, Feynman's Rainbow,* and *Euclid's Window.* His scientific writings have appeared in journals from *Nature* to *Scientific American,* and he has also written for *The New York Times, The Wall Street Journal, The New York Review of Books, Wired,* and *Psychology Today,* as well as for the television series *MacGyver* and *Star Trek: The Next Generation.*

A NOTE ON THE TYPE

The text of this book was set in a typeface called Aldus, designed by the celebrated typographer Hermann Zapf in 1952–53. Based on the classical proportions of the popular Palatino type family, Aldus was originally adapted for Linotype composition as a slightly lighter version that would read better in smaller sizes.

Hermann Zapf was born in Nuremberg, Germany, in 1918. He has created many other well-known typefaces, including Comenius, Hunt Roman, Marconi, Melior, Michelangelo, Optima, Saphir, Sistina, Zapf Book, and Zapf Chancery.

Composed by North Market Street Graphics,
Lancaster, Pennsylvania

Printed and bound by Berryville Graphics,
Berryville, Virginia

Designed by Cassandra J. Pappas